Lecture Notes in Mathematics

Edited by A. Dold and B. Eckmann

Series: Mathematisches Institut der
Universität Erlangen-Nürnberg
Advisers: H. Bauer and K. Jakobs

816

Lucretiu Stoica

Local Operators and Markov Processes

Springer-Verlag
Berlin Heidelberg New York 1980

Author

Lucretiu Stoica
Department of Mathematics, INCREST
Bdul. Pacii 220
77538 Bucharest/Romania

AMS Subject Classifications (1980): 31D05, 60J25, 60J35, 60J40, 60J45

ISBN 3-540-10028-8 Springer-Verlag Berlin Heidelberg New York
ISBN 0-387-10028-8 Springer-Verlag New York Heidelberg Berlin

Library of Congress Cataloging in Publication Data. Stoica, Lucretiu, 1949- Local operators and Markov processes. (Lecture notes in mathematics; 816) Bibliography: p. Includes index. 1. Markov processes. 2. Operator theory. 3. Potential, Theory of. I. Title. II. Series: Lecture notes in mathematics (Berlin); 816.
QA3.L28 no. 816 [QA274.7] 510s [519.2'33] 80-20217

Printing and binding: Beltz Offsetdruck, Hemsbach/Bergstr.
2141/3140-543210

Introduction

The present book deals with the axiomatic potential theory of local character and its corresponding continuous standard processes.This subject has already been treated by several authors.

P.A. Meyer [26] constructed a Hunt process on a Brelot space such that the class of all excessive functions and the class of all positive hyperharmonic functions should coincide. This result was generalized and completed by N.Boboc, C.Constantinescu and A.Cornea [8], W.Hansen [19], I.Cuculescu [16], H.Bauer [2], C.Constantinescu and A.Cornea [13]. The converse problem, the construction of an axiomatic potential theoretic structure associated to a given continuous Markov process, was first approached by Ph. Courrège and P.Priouret [14]. Then J.C. Taylor [40] and J.Bliedtner and W.Hansen [5] proved that each continuous standard process, whose potential kernel is strong Feller, yields a harmonic space, in the meaning of C.Constantinescu and A.Cornea [13].

The classical examples of the axiomatic potential theory of local type and of continuous Markov processes are associated to second order elliptic or hypoelliptic differential operators. On a locally compact space a fairly good substitute for the differential operators are the local operators. The notion of a local operator was introduced by E.B. Dynkin [17] p.145. He associated a local operator to each continuous standard process. The relation between this notion and the kernels from the axiomatic potential theory was pointed out by G.Mokobodzki and D.Sibony [31] Th. 21. Further boundary value problems associated to local operators were considered by G.Lumer [23], [24] and J.P. Roth [34].

The main axiomatic potential theoretic object studied here is a local operator, L, on a locally compact space, X, with a countable base. Similar to elliptic differential operators our local operator is assumed to obey a maximum principle and to have a base of open sets that are regular for the Poisson-Dirichlet problem. We note that G.Lumer (in [23]) was the first to consider a similar framework, but the spirit of the present approach differs much from his.

In Chapter I we construct a continuous standard process with state space X such that its characteristic operator extends L and its transition function is unique. Then we characterize the excessive functions of the process by means of the Dirichlet problem for L.

The sheaf of all excessive functions and the hitting distributions (or harmonic measures) may be viewed as the invariants under the random time change transformations. In Chapter II we suggest an axiomatic approach for these objects, taking as axioms some of the properties proved in the first chapter. First we construct the potential kernel associated to a potential (function) by computing it in terms of the given harmonic measures. In Section 6. of Chapter II we construct an open covering $\{U_i | i \in I\}$ and for each i a continuous strict potential, p_i, on U_i such that $p_i - p_j$ is harmonic on $U_i \cap U_j$ for $i \neq j$. This is somewhat analogous with the problem of constructing the random time change for two processes with identical hitting distributions. By analogy an inequality similar to the one proved by R.M. Blumenthal and R.K. Getoor [6] turns out to be important. This inequality is proved in Section 3. From it we deduce some remarkable properties of the potential kernel associated to a continuous strict potential. Finally a local operator possessing the properties considered in Chapter I is associated to the family $\{U_i , p_i\}$.

In Chapter III we study the topological properties of the transition function of the process constructed in Chapter I. Using the results of Chapter II we show that the transition function maps the cone of all lower semicontinuous functions into itself and the range of the resolvent has a "local density" property. If X is compact and if there is a function $h>0$ such that $Lh=0$, then the transition function maps the space of all continuous functions into itself.

The study of a product space is a classical theme in potential theory. While the early papers (K. Gowrisankaran [18], R. Cairoli [12]) study functions on the product space which are related to the structures of the terms of the product, in Chapter IV of the present work we follow the idea of the probabilistic work of R. Cairoli [11], constructing a structure on the product space and studying this structure. Namely we construct local operators on product spaces. This subject is a particular aspect of the general problem of constructing the notion of product in potential theory (a problem suggested by N.Boboc).

In Chapter IV we first consider two local operators L^1, L^2 on locally compact spaces X_1 , X_2 which possess bases of regular sets. Then we construct the sum L^1+L^2 on $X_1 \times X_2$ and prove that the product of two regular sets is regular (for L^1+L^2). Then we prove a similar result for the sum of a series of local operators on the product of a sequence of compact spaces. Further we consider a local operator, L, and construct the operator L-d/dt. (A similar construction within a different framework was made by J.P. Roth [34]). Then we are interested in those local operators which yield Bauer spaces and the operator L-d/dt allows us to characterize those operators with the property that, by addition they also yield Bauer spaces on product spaces. (The problem was also treated by E.Popa [33] and U. Schirmeier [36] in the frame of harmonic spaces in the sense of C.Constantinescu and A.Cornea. The key technical result is Lemma 5.5). Finally it is shown that the sum of a series of local operators preserves these properties under suitable conditions. This result extends a (more precise) result of C.Berg to compact spaces. (He constructed a Brelot space on the infinite dimensional torus [3]).

In Section 1. of Chapter V we consider the case when the state space, X, is a locally compact abelian group. We show that for a given translation invariant structure of the type considered in Chapter II, there exists a unique translation invariant local operator associated to it. In Section 2. we show that local operators can be used on a harmonic space in the sense of Constantinescu and Cornea (althogh there is no base of regular sets) and all the results from the previous sections rest valid in a natural analogous form.

Chapter VI is devoted to Feller resolvents. In Section 1. we present an improvement of a wellknown result on convex cones of lower semicontinuous functions. In Section 2. we give a very general construction of Hunt processes. Section 3. contains an excessiveness criterion. Section 4. presents a characterisation of those Feller resolvents which yield continuous Hunt processes (Corollary 4.12).

By analogy with the study made in Section 5. of Chapter IV, the final note gives a characterisation for the semigroups of compact contractions in Hilbert spaces.

Most of the material in this book was previously presented at the Potential Theory Seminar in Bucharest.

I would like to express my thanks to professors N.Boboc, Gh. Bucur, A.Cornea and I.Cuculescu from whom I learned potential theory and Markov Processes.

The expert typing was done by Camelia Minculescu, to whom I want to express my gratitude.

C O N T E N T S

NOTATION

For a locally compact space with a countable base, T, we shall denote by $C(T)$ the space of all real continuous functions and by $C_o(T)$, $C_c(T)$, $C_b(T)$ the subspaces of functions vannishing to infinity, of functions of compact support, of bounded functions. The space of all real Borel functions on T will be denoted by $\mathcal{B}(T)$ and the subspace of bounded Borel functions by $\mathcal{B}_b(T)$.

A kernel on T will be a positive linear operator V from $\mathcal{B}_b(T)$ into $\mathcal{B}(T)$ such that for each $x \in T$ the map $f \longrightarrow Vf(x)$ defines a Radon measure. The measure associated to x is denoted by V^x, i.e. $V^x(f)=$ $=Vf(x)$.

All terminology and notation on Markov processes will be that of [6]. Particularly if $(\Omega,M,M_t,X_t,\theta_t,P^x)$ is a standard process with state space $(E,\mathcal{E})$, f is a nearly Borel positive function and A is a nearly Borel set we use the notation $T_A=\inf\{t>0 / X_t \in A\}$, $P_A^{\lambda}f(x)=$ $=E^x[\exp(-\lambda T_A).f(X_{T_A}), T_A<\infty]$ and $P_A=P_A^0$.

We say that a standard process is continuous if $t \longrightarrow X_t$ is a.s. continuous on $[0,\zeta)$.

For the terminology and notation from the theory of harmonic spaces which is not specifically explained here we refer to [13].

I. LOCAL OPERATORS

1. General Properties

1.1. A sheaf of vector spaces of real continuous functions on a locally compact space, X, is a family $\{A(U)/U$ open set$\}$ such that:

1° For each open set U, $A(U)$ is a vector space of real continuous functions on U;

2° If $U_1 \subset U_2$ are open sets and $f \in A(U_2)$ then $f_{|U_1} \in A(U_1)$;

3° If $\{U_i / i \in I\}$ is a family of open sets, $U=\bigcup_{i\in I} U_i$ and $f \in C(U)$ satisfies $f_{|U_i} \in A(U_i)$, then $f \in A(U)$.

1.2. A local operator, L, on a locally compact space, X, is a pair $(\{D(U,L)/U$ open set$\}$, $\{(L,U)/U$ open set$\})$, where $\{D(U,L)/U$ open set$\}$ is a sheaf of vector spaces of real continuous functions on X and $\{(L,U)/U$ open set$\}$ is a family of linear operators such that:

1° $(L,U) : D(U,L) \longrightarrow C(U)$ is a linear operator.

2° If U, V are open sets, $U \subset V$ and $f \in D(V,L)$ then $(L,U)(f_{|U}) = ((L,V)f)_{|U}$ (i.e. L is a sheaf morphism from $\{D(U,L)/U$ open set$\}$ into the sheaf of all continuous functions).

We shall use the notation $(L,U)f = Lf$ for any open set U and any $f \in D(U,L)$ (just like in the case of differential operators in R^n).

1.3. <u>In this section</u> we shall consider a locally compact space with a countable base, X, and a local operator, L, on X. For each $\lambda > 0$ we denote by L_λ the operator defined as follows: $D(U,L_\lambda) = D(U,L)$ and $L_\lambda f = Lf - \lambda f$ for any any open set, U, and any $f \in D(U,L)$.

1.4. Suppose that U is a relatively compact open set such that $\partial U \neq \Phi$.

U will be called Dirichlet regular (or D-regular) if:

1° $(\forall)$ $f \in C(\partial U)$, $(\exists)$ $u \in C(U)$ unique such that $u_{|\partial U} = f$, $u_{|U} \in D(U,L)$, and $Lu = 0$ on U,

2° if $f \geq 0$, then the associated function, u, satisfies $u \geq 0$.

If U is D-regular and $f \in C(\overline{U})$ we shall denote by $H^U f = u$, the function associated to $f_{|\partial U}$, via 1° in the above definition. H^U may be regarded as a linear operator, $H^U : C(\overline{U}) \longrightarrow C(\overline{U})$, which extends to a kernel on $\overline{U}$. If U is D-regular with respect to L_λ, then we shall use the notation H^U_λ for the analogous object.

U will be called Poisson regular (or P-regular) if:

1° $(\forall)$ $f \in C(\overline{U})$, $(\exists)$ $u \in C_o(\overline{U}) \cap D(U,L)$ unique such, that $Lu = -f$,

2° if $f \geq 0$, then the associated function, u, satisfies $u \geq 0$,

3° the space $D_o(U)$ is dense in $C_o(U)$, where

(1) $$D_o(U) = f \in \{D(U,L) \cap C_o(U)/Lf \in C_o(U)\}.$$

If U is both P-regular and D-regular we shall call it P-and D-regular.

If U is P-regular and $f \in C(\overline{U})$ we shall denote by $G^U f = u$, the function associated to f via 1° in this definition. This way we get a positive linear operator G^U: $C(\overline{U}) \longrightarrow C(\overline{U})$, which extends to a kernel on $\overline{U}$. Condition 3° shows that for any $x \in U$ the measure $G^{U,x}$ is nonnul, and hence $G^U 1 > 0$ on U. If U is P-regular with respect to L_λ, then G^U_λ will denote the analogous object (of course $G^U_o = G^U$). If U is P-regular with respect to L_λ for any $\lambda > 0$ and $f \in C(U)$, then

$$L(G^U_\alpha f - G^U_\beta f) - \alpha(G^U_\alpha f - G^U_\beta f) = (\alpha - \beta) G^U_\beta f \quad ,$$

which leads to $G^U_\alpha - G^U_\beta = (\beta - \alpha) G^U_\alpha G^U_\beta$, $\alpha, \beta > 0$, i.e. the rezolvent equation.

1.5. The operator L will be called locally closed (24(II) p.207) if:

$(\forall)$ U open set, $(\forall)$ $\{f_n/n\in N\}\subset D(U,L)$, $f_n \longrightarrow f$, $Lf_n \longrightarrow g$

uniformly on each compact set

$$\Longrightarrow f\in D(U,L) \quad \text{and } Lf=g.$$

We remark that L is locally closed provided there exists a base of open sets which are P- and D-regular. In order to see this we consider an open set, U, and a P- and D-regular set, V, such that $\overline{V}\subset U$; then we deduce

$$\varphi=H^V_\varphi+G^V(-L\varphi) \quad \text{on } V, \qquad (\forall)\ \varphi\in D(U,L). \tag{2}$$

Writting this formula for the sequence $\{f_n\}$ and letting $n\longrightarrow\infty$ we get $f=H^Vf+G^V(-g)$, which shows $f\in D(V,L)$ and $Lf=g$ on V.

1.6. The operator L will be called locally dissipative if it obeys the following maximum principle:

$(\forall)$ U open set, $(\forall)$ $f\in D(U,L)$, $(\forall)$ $x\in U$,

$f(x)\geqslant f$ on U, $f(x)\geqslant 0 \Longrightarrow Lf(x)\leqslant 0$.

<u>From now on we suppose that L is locally dissipative</u>. Then L_λ, $\lambda>0$ are locally dissipative. First we are going to state a very useful form of the minimum principle. Versions of it were proved in several places (see [17 (I)] p.145 and [24](II) p.210).

1.7. <u>Proposition</u>

Suppose that U is an open set and $\varphi\in D(U,L)$ satisfies $|\varphi|\leqslant 1$ and $L\varphi<0$. If $f\in D(U,L)$, $Lf\geqslant 0$ and $\limsup_{x\to\infty(u)} f(x)\leqslant a$, $a\in R_+$, where $\infty(U)$ is the Alexandrov point associated to the locally compact space U, then $f\leqslant a$ on U.

<u>Proof</u>

Let us suppose that $f(x)\geqslant a+\alpha$, $\alpha>0$ for some $x\in U$. Then $f-(\alpha/2)\varphi=g$ verifies

$$g(x)\geqslant a+\alpha/2 \quad \text{and } \limsup_{y\to\infty(U)} g(y)\leqslant a+\alpha/2.$$

There exists a maximum point, $y\in U$, such that $g(y)\geqslant g$ on U. Then $g(y)\geqslant g(x)\geqslant 0$, and hence $Lg(y)\leqslant 0$. On the other hand

$$Lg(y)=Lf(y)-(\alpha/2)L\varphi(y)>0,$$

which is a contradiction. Our supposition failed, and hence $f(x)\leqslant a$ for any $x\in U$.

Now we are going to introduce the "local closure" of L. First we need a version of a result from [34] p.55.

1.8. Proposition

Let V be an open set, $\{f_n/n\in N\}\subset D(V,L)$ a sequence such that $f_n \longrightarrow f$, $Lf_n \longrightarrow \psi$ uniformly on each compact set, and f has nonnegative local maximum in $x_o\in V$. Assume that for any neighbourhood, W, of x_o there exist an open set, U, such that $x_o\in U$, $\overline{U}\subset W$ and $g\in C_o(U)\cap D(U,L)$ such that $g(x_o)>0$, $Lg\in C_b(U)$. Then $\psi(x_o)\leqslant 0$.

Proof

Let us suppose that $\psi(x_o)>0$. We choose an open set, U, such that $\overline{U}\subset V$, $x_o\in U$, $\psi>\alpha$ on U, $\alpha\in R$, $\alpha>0$, $f(x_o)\geqslant f$ on U and $g\in C_o(U)\cap D(U,L)$ such that $g(x_o)=\beta>0$, $|Lg|\leqslant\alpha/2$. Further we choose $n\in N$ such that $|f_n-f|<\beta/2$ on $\overline{U}$ and $|Lf_n-\psi|<\alpha/2$ on $\overline{U}$. Then we have

$$f_n(x_o)+g(x_o)>f(x_o)+\beta/2 \;,$$

$$f_n(y)+g(y)=f_n(y)<f(y)+\beta/2\leqslant f(x_o)+\beta/2 \;,\quad (\forall)\; y\in\partial U$$

and

$$Lf_n+Lg=Lf_n-\psi+\psi+Lg>-\alpha/2+\alpha-\alpha/2=0 \;.$$

This contradicts 1.7, and hence $\psi(x_o)\leqslant 0$.

1.9. Corollary

Let us assume that the following condition holds:

$(\forall)$ $x\in X$, $(\forall)$ V a neighbourhood of x, $(\exists)$ U open set, $x\in U$, $\overline{U}\subset V$, $(\exists)$ $g\in C_o(U)\cap D(U,L)$ such that $g(x)>0$ and $Lg\in C_b(U)$.

Then for each open set, V, and each sequence, $\{f_n\}\subset D(V,L)$, such that $f_n\longrightarrow 0$ and $Lf_n\longrightarrow\psi$ uniformly on the compact subsets of V it holds $\psi\equiv 0$.

If the requirement from the above corollary is fulfilled, we may define $\tilde{L}$, the local closure of L, as follows:

If U is an open set, a function $f\in C(U)$ belongs to $D(U,\tilde{L})$ if and only if there exist a function $\psi\in C(\overline{U})$, an open covering of U, $\{U_i/i\in I\}$, and for any $i\in I$ there exists a sequence $\{\varphi_n^i/n\in N\}\subset D(U_i,L)$ such that $\varphi_n^i\longrightarrow f$ and $L\varphi_n^i\longrightarrow\psi\,(n\longrightarrow\infty)$ uniformly on the compact subsets of U_i. Furthermore we put $\tilde{L}f=\psi$.

We note that from 1.8 one deduces $\tilde{L}$ is also locally dissipative.

The next proposition is a criterion of P-regularity and also shows that the kernel G^U is supported by U.

1.10. Proposition

Suppose that U is an open set such that $\partial U\neq\emptyset$ and for any

$f \in C(\overline{U})$ there exists a function $u \in C_o(U) \cap D(U,L)$ which fulfils $Lu=-f$ and the space $C_o(U) \cap D(U,L)$ is dense in $C_o(U)$. Then U is P-regular and $G^U(\partial U)=0$.

Proof

Let $u \in C_o(U) \cap D(U,L)$ be such that $Lu=-f$, $f \in C(U)$, $f \geqslant 0$. Proposition 1.7 implies $u \leqslant 0$. Thus 2° and the unicity assertion of 1° within the definition of P-regularity are fulfilled. Further the operator G^U exists and may be extended to a kernel on U. Next we are going to prove $G(\partial U)=0$. Let $\{\phi_n\} \subset C(U)$ be a sequence such that $0 \leqslant \phi_{n+1} \leqslant \phi_n \leqslant 1$, $\overline{\{\phi_n < 1\}} \subset U$, and $\bigcup_n \overset{o}{\overbrace{\{\phi_n=0\}}} = U$. From 1.7 we get

$$\|G\phi_n\| \leqslant \sup \{G\phi_n(x)/\phi_n(x) > 0\} \leqslant \sup \{G1(x)/\phi_n(x) > 0\} .$$

But $G1 \in C_o(U)$ and so $G(\partial U) = \lim_{n \to \infty} G(\phi_n)=0$.

Now let $u \in C_o(U) \cap D(U,L)$, $Lu \in C(\overline{U})$, $Lu \leqslant 0$. For $x \in U$ we have $u(x)=G(-Lu)(x) = \lim_{n\to\infty} G((-Lu)(1-\phi_n))(x)$. The limit being increasing it is uniforme. This leads to condition 3° from the definition of P-regularity.

1.11. The next theorem due to G.A. Hunt will be used several times in our paper. For a proof we refer to [30] p.223-224 or [27] X T 10. An extend study of this subject can be found in [22].

Theorem

Let $V : C_b(X) \longrightarrow C_b(X)$ be a positive linear operator satisfying the complete maximum principle, i.e.:

if $f,g \in C_{b+}(X)$, $Vf(x) \leqslant Vg(x)+1$, $(\forall)\ x \in \{f>0\}$,

then $Vf \leqslant Vg+1$.

Then there exists a unique family $\{V_\lambda/\lambda \geqslant 0\}$ of positive linear operators on $C_b(X)$ such that

1° $V_\alpha - V_\beta = (\beta-\alpha)V_\alpha V_\beta$, $\alpha,\beta \geq 0$,

2° $\lambda V_\lambda 1 \leq 1$, $\lambda>0$,

3° $V_o = V$.

1.12. Now we are going to depict several relations between the various kinds of regularity.

1° If an open set, U, is P-regular, then it is P-regular with respect to L_λ for any $\lambda>0$ and the resolvent $\{G^U_\lambda/\lambda>0\}$ is sub-Markov: $\lambda G^U_\lambda 1 \leq 1$.

This is a consequence of the above theorem applied to the operator G^U. The complete maximum principle for G^U results from 1.7.

2^0 If U is P-regular with respect to L_α, for some $\alpha>0$, then U is P-regular with respect to L_λ for any $\lambda>0$ and the resolvent $\{G_\lambda^U/\lambda>0\}$ is sub-Markov.

This results from Theorem III.3.1 of F.Hirsch [22].

3^0 If U is P-regular with respect to L_λ for any $\lambda>0$ then U is P-regular (with respect to L) provided there exists $f\in C_b(U)\cap D(U,L)$ such that $Lf\in C_b(U)$ and $Lf\leq -1$.

In order to prove this we choose $\alpha>0$ such that $\alpha||f||<1/2$ and put $\varphi=2G_\alpha^U(-L_\alpha f)$. Then

$$L\varphi=2(Lf-\alpha(f+G_\alpha^U L_\alpha f)) .$$

Using 1.7 we get $|f+G_\alpha^U L_\alpha f|\leq||f||$, and hence $L\varphi\leq -1$, which leads to $G_\lambda^U 1\leq\varphi$ for any $\lambda>0$. Now using the kernel $G^U=\lim_{\lambda\to 0} G_\lambda^U$ it is easy to deduce that U is P-regular.

4^0 Let U be D-regular and suppose there exists $\lambda>0$ such that U is also P-regular with respect to L_λ. Then U is D-regular with respect to L_λ and

(3) $$H_\lambda^U=H^U-\lambda G_\lambda^U H^U.$$

5^0 Let U be an open set. Assume that L is locally closed and there exists a P-regular set V such that $\bar{U}\subset V$. If U is P-regular, then it is also D-regular and for any $f\in C_b(V)$,

(4) $$G^V f-G^U f=H^U G^V f \qquad \text{on } U.$$

Condition 2^0 and the unicity assertion from 1^0 within the definition of D-regularity are consequences of 1.7. In order to prove the existence assertion we firstly consider the case when $f\in C(\partial U)$ is of the form $f=G^V g_{|\partial U}$ for suitable $f\in C_b(V)$. Then $u=G_g^V-G_g^U$ fulfils $u_{|\partial U}=f$ and $Lu=0$ on U. For a general $f\in C(\partial U)$ one makes an approximation.

6^0 Let U be an open set and V a P-regular set such that $\bar{U}\subset V$. If U is D-regular then it is also P-regular. The proof of this assertion is similar with the preceding one.

2. The Markov Process Associated to a Local Operator

Let X be a locally compact space with a countable base. In this section we study a local operator on X, L, which is locally dissipative and suppose that the family of all P- and D-regular sets forms a topological base. Then from 1.12, 1° we deduce similar properties with respect to the operators L_{λ} , $\lambda>0$.

2.1. Proposition

a) Let V be a P-regular set. There exists a continuous Hunt process $(\Omega, M, M_t, X_t, \theta_t, P^x)$ with state space V such that for each $\varphi \in C_o(V)$ and $t>0$ the function $\psi(x)=E^x[\varphi(X_t)]$ satisfies $\psi \in C_o(V)$ and

$$(1) \qquad G_{\lambda}^{V} f(x)=E^x\left[\int_0^{\infty} \exp(-\lambda t) f(X_t)\,dt\right], \quad (\forall)\ x \in V, \quad \lambda \geq 0,$$

$$f \in C_b(V).$$

b) If U is a D-regular set, $\overline{U} \subset V$, then each point $x_o \in \partial U$ is regular, i.e. $E^{x_o}[T_{V\setminus U}>0]=0$, and the following equalities hold:

$$(2) \quad H_{\lambda}^{U} f(x)=P_{V\setminus U}^{\lambda} f(x), \quad (\forall)\ x \in \overline{U}, \quad \lambda \geq 0, \quad f \in C(V),$$

$$(3) \quad G_{\lambda}^{U} f(x)=E^x\left[\int_0^{T_{V\setminus U}} \exp(-\lambda t) f(X_t)\,dt\right], \quad (\forall)\ x \in U,\ \lambda \geq 0,\ f \in C_b(U).$$

In order to prove this proposition we need the next three lemmas:

2.2. Lemma. Let $(\Omega, M, M_t, X_t, \theta_t, P^x)$ be a standard process with state space E. Assume that there exists a sequence $\{B_n\}$ of nearly Borel sets such that $\bigcup_n B_n=E$ and $R(B_n)(x)=E^x[\int_0^{\infty} \chi_{B_n}(X_t)\,dt]$, $(x \in E)$ is a bounded function for each $n \in N$. Suppose that A is a nearly Borel set and H is a kernel on E such that

1° $H(E\setminus A)=0$,

2° $Hf \leq f$ for any excessive function f,

3° there exists a family $\mathcal{A}$ of excessive functions such that: (a) any two measures on E, μ and υ, coincide provided $\mu(f)=\upsilon(f)$ for any $f \in \mathcal{A}$; (b) Hf is excessive for any $f \in \mathcal{A}$ and $Hf=f$ on A.

Then $P_A=H$ and all points in A are regular.

Proof. Let $f \in \mathcal{A}$, and g be an excessive function such that $f \leq g$ on A; them from 1° and 2° we get $Hf \leq Hg \leq g$ and on account of 3°(b) deduce

$$Hf=\inf\{\, g/\text{excessive},\ f \leq g \text{ on } A\}.$$

On the other hand Hunt's balayage theorem ([6] p.141) gives us $P_A f \leq Hf$ and $P_A f(x)=Hf(x)$ except possibly for those points, x, in A which are not regular, i.e. except for a semipolar set ([6] p.80). But $P_A f$ is also excessive ([6] p.73), hence $Hf=P_A f$. Now condition 3°(a) implies

$H=P_A$.

If $x\in A$ then conditions 3^o(a), (b) show that $H^x=\varepsilon_x$. Thus $R(B_n)(x)=E^x[R(B_n)(X_{T_A})]$ or $E^x[\int_0^\infty \chi_{B_n}(X_t)dt]=E^x[\int_{T_A}^\infty \chi_{B_n}(X_t)dt]$ for any $n\in N$. We deduce $E^x[\int_0^{T_A} \chi_{B_n}(X_t)dt]=0$ for any $n\in N$, and hence $E^x[T_A>0]=0$.

2.3. Lemma. Let $g\in C_{b+}(V)$ and put

$$h(x)=\begin{cases} g(x) & \text{if } x\in V\setminus U \\ H^U g(x) & \text{if } x\in U\end{cases}$$

Then k is excessive for the resolvent $\{G_\lambda^V \mid \lambda\geq 0\}$, i.e. $\lambda G_\lambda^V h\to h$, as $\lambda\to\infty$, and $\lambda G_\lambda^V h\leq h$, for $\lambda>0$.

Proof. Since V is P-regular we know that $\overline{G_\lambda^V(C_0(V))}=C_0(V)$ for for each $\lambda\geq 0$. Therefore $\lambda G_\lambda^V f\to f$, as $\lambda\to\infty$, for each $f\in C_0(V)$. Since $h\in C_0(V)$ we have only to prove the inequality $\lambda G_\lambda^V h\leq h$. From 1.7 we get $h\leq g$ and $\lambda G_\lambda^V h\leq \lambda G_\lambda^V g\leq g=h$ on $V\setminus U$. On the other hand $L_\lambda(\lambda G_\lambda^V h-h)=0$ on U. Again 1.7 gives us $\lambda G_\lambda^V h\leq h$ on U.

The next Lemma was proved by Ph.Courrège and P.Priouret in Annexe 1 of [14].

2.4. Lemma. Let $(\Omega, M, M_t, X_t, \theta_t, P^x)$ be a standard process with state space E. If there exists a base of open sets, $\mathcal{U}$, such that

$$P_{CU}(E\setminus\overline{U})(x)=0, \quad \text{for each } U\in\mathcal{U} \text{ and each } x\in U,$$

then the process is continuous.

Proof of Proposition2.1. a) The resolvent $\{G_\lambda^V \mid \lambda>0\}$ satisfies the conditions from the Hille-Yosida theorem, on the Banach space $C_0(V)$. Thus we get a (C_0)-class semigroup of positive sub-Markov operators on $C_0(V)$. Further we apply the theorem from [6] p.46 and get a standard process $(\Omega, M, M_t, X_t, \theta_t, P^x)$ with state space V which fulfils relation (1). The continuity of the process results from Lemma 2.4 by using relation (2), which will be proved below.

b) In order to prove (2) we are going to apply Lemma 2.2 with respect to the kernel H^U (extended to V by taking $H^{U,x}=\varepsilon x$ for $x\in V\setminus U$), the set $A=V\setminus U$ and the family $A=G^V(C_{b+}(V))$.

Conditions 1° and 3° (a) from Lemma 2.2 are obviously fulfilled. From 1.7 we get $H^U G^V g \leq G^V g$ for each $g \in C_{b+}(V)$. Then the monotone class theorem shows that this inequality is still valid for each $f \in B_{b+}(V)$. Further we get $H^U f \leq f$ for each excessive function, by approximating f with potentials. This cheks condition 2° from 2.2. Condition 3^o(b) 2.2 results from Lemma 2.3. Thus relation (2) follows from Lemma 2.2.

Now let $f \in C_b(V)$. The strong Markov property gives us

$$H^U G^V f(x) = E^x[E^{X_{T_{V\setminus U}}}[\int_0^\infty f(X_t)\,dt\,]] = E^x[\int_{T_{V\setminus U}}^\infty f(X_t)\,dt\,].$$

This relation together with 1(4) leads to (3).

2.5. Theorem. There exists a continuous standard process $(\Omega, M, M_t, X_t, \theta_t, P^x)$ with state space X such that for any P-regular set, U,

$$(4) \qquad G^U_\lambda f(x) = E^x[\int_0^{T_{E\setminus U}} \exp(-\lambda t) f(X_t)\,dt], \qquad (\forall) \quad x \in U,$$

$$(\forall) \ f \in C_b(U) \ , \quad (\forall) \ \lambda \geq 0.$$

If another continuous standard process fulfils (4), then it has the same transition function.

This theorem is a consequence of the next theorem proved by Ph.Courrège and P.Priouret [15] 2.4.2. (See also P.A. Meyer [42] and M.Nagasawa [43]).

2.6. Theorem. Let E be a topological space which is homeomorphic with a Borel set from a compact metric space. Let $\{E_i/i\in I\}$ be an open covering of E and for each $i\in I$ let $(\ldots,X_t^i,\ldots)$ be a continuous strong Markov process on E_i. Assume that for each pair (i,j) such that $E_i\cap E_j\neq\emptyset$ both processes $(\ldots,X_t^i,\ldots)$ and $(\ldots,X_t^j,\ldots)$ induce (by killing) the same transition function on $E_i\cap E_j$. Then there exists a continuous strong Markov process on E which for each $i\in I$ induces on E_i (by killing) a transition function identical with that of $(\ldots,X_t^i,\ldots)$. The transition function of such a process on E is unique.

Proof of Theorem 2.5. Let V_1, V_2 be two P-regular sets and U a D-regular one such that $\overline{U}\subset V_1\cap V_2$. Proposition 2.1 gives us a continuous standard process on V_1 and another continuous process on V_2. From (3) we deduce that both processes yield (by killing on CU) the same transition function on U; namely the Laplace transform of this transition function is G_λ^U, $\lambda\geq 0$. Therefore the unicity assertion from 2.6 allows us to deduce that both processes yield (by killing on $C(V_1\cap V_2)$) the same transition function on $V_1\cap V_2$. Further 2.6 proves the existence of a continuous process on X that fulfils (4) and the unicity of its transition function.

Now let us recall the notion of a characteristic operator of continuous process. Let $(\Omega,M,M_t,X_t,\theta_t,P^x)$ be a continuous standard process with state space E. Let $x\in E$ and f a universally measurable function defined on an open set U with $x\in U$. We denote by $F_f(x)$ the family of all open sets, V, such that $\overline{V}\subset U$, $x\in V$ and $P_{CV}f(x)<\infty$ and consider it as a directed set. Further $\mathcal{D}(x)$ will be the family of all universally measurable functions f, defined in an open neighbourhood of x such that $F_f(x)\searrow\{x\}$ and $\lim\limits_{V\searrow\{x\}}(P_{CV}f(x)-f(x))/E^x[T_{CV}]$ exists, where the limit is taken over $F_f(x)$. Then the characteristic operator of E.B. Dynkin (17 I p.140 and p.145) is denoted by $\mathcal{U}$. For $x\in E$ and $f\in\mathcal{D}(x)$,

$$\mathcal{U}f(x)=\lim_{\substack{V\searrow\{x\}\\ V\in F_f(x)}}(P_{CV}f(x)-f(x))/E^x[T_{CV}]\ .$$

Now we remark that the characteristic operator, $\mathcal{U}$, of the process constructed in Theorem 2.5 extends L, i.e.

$(\forall)$ U open set, $(\forall)\, f\in D(U,L)$ $(\forall)\, x\in U$
$f\in\mathcal{D}(x)$ and $\mathcal{U}f(x)=Lf(x)$.

This property results from Lemma 5.7 of E.B. Dynkin [17].

In chapter III we shall show that the process constructed in Theorem 2.5 is a Hunt process. Thus it is reasonable to prove its unicity within the class of all Hunt processes:

2.7. Corollary. If $(\Omega,\mathcal{M},\mathcal{M}_t,X_t,\theta_t,P^x)$ is a continuous Hunt process whose characteristic operator, $\mathcal{U}$, extends L, then it has the same transition function as the process constructed in Theorem 2.5.

Proof. Let U be a P-regular set and $x\in U$. For $f\in C_b(U)$ and $0<\varepsilon<1$ we denote by $\Lambda=\Lambda(x,U,f,\varepsilon)$ the family of all stopping times, T, such that $T\le T_{CU}$, a.s., $E^x[T]<\infty$ and

$$(5)\qquad |G^Uf(x)-E^x[G^Uf(X_T)]-E^x[\int_0^T f(X_t)]|\le\varepsilon E^x[T]\ ,$$

$$|G^U1(x)-E^x\, G^U1(X_T)\ -E^x[T]|\le\varepsilon E^x[T]\ .$$

This family has the following properties:

a) If $\{T_n\}$ is a sequence from Λ and $T_n\le T_{n+1}$, P^x-a.s. for each n, then $\limsup_{n\to\infty} T_n$ belongs to Λ.

b) If $T\in\Lambda$ and $P^x(T<T_{CU})>0$, then there exists $T'\in\Lambda$ such that $T\le T'$ and $P^x(T<T')>0$.

In order to check property a) one uses the inequality $(1-\varepsilon)E^x[T]\le G^U1(x)$ deduced from (5), which shows $E^x[\lim_{n\to\infty} T_n]<\infty$. Let us check property b). First we remark that for each $y\in X$ there exists an open set, V, such that $y\in V$ and $E^y[T_{CV}]<\infty$, because $\mathcal{U}f(y)\ne0$ for some $f\in\mathcal{D}(y)$. If $T\in\Lambda$ satisfies $P^x(T<T_{CU})>0$, then we can choose an open set V and a Borel subset K such that $K\subset V$, $V\subset U$, $P^x(X_T\in K)>0$ and $E^y[T_{CV}]\le1$ for each $y\in K$. Since $\mathcal{U}G^Uf(y)=-f(y)$ for each $y\in U$, we can choose another open set, V', and another Borel set, K', such that $V'\subset V$, $K'\subset K\cap V'$, $P^x(X_T\in K')>0$ and

$$|G^Uf(y)-E^y[G^Uf(X_{T_{CV'}})]-f(y)E^y[T_{CV'}]|\le(\varepsilon/2)E^y[T_{CV'}]\ ,$$

$$|G^U1(y)-E^y[G^U1(X_{T_{CV'}})]-E^y[T_{CV'}]|\le(\varepsilon/2)E^y[T_{CV'}]\ ,$$

for each $y\in K'$,

$$|f(y)-f(z)|\le\varepsilon/2\quad\text{for each } y,z\in V'\ .$$

Then, from these relations we deduce

$$|G^Uf(y)-E^y[G^Uf(X_{T_{CV'}})]-E^y[\int_0^{T_{CV'}} f(X_t)\,dt]|\le\varepsilon E^y[T_{CV'}]\ .$$

Then putting $T'(\omega)=T(\omega)+T_{CV'}\circ\theta_T(\omega)$ if $X_T(\omega)\in K'$ and $T'(\omega)=T(\omega)$ if $X_T(\omega)\notin K'$ we deduce relation (5) with T' instead of T.

On the other hand $E^x[T'] = E^x[T] + E^x[E^{X_T}[T_{CV'}]/X_T \in K'] < \infty$, and hence $T' \in \Lambda$. Since $P^x(T'>T) = P^x(X_T \in K') > 0$, we have proved property b).

Now we consider Λ as an ordered set with the order defined by "$T \leq T'$ iff $P^x(T'<T)=0$". From property a) one deduces that Λ is inductive ordered and Zorn's Lemma gets us a maximal element $T \in \Lambda$. From property b) we deduce $T=T_{CU}$, P^x- a.s. Thus relation (5) with $T=T_{CU}$ becomes

$$|G^U f(x) - E^x[\int_0^T f(X_t)dt]| \leq \varepsilon E^x[T_{CU}] \quad .$$

ε being arbitrary we get $G^U f(x) = E^x[\int_0^T f(X_t)dt]$. Now the assertion results from Theorem 2.5.

3. Elliptic degenerated operators in R^n

In this section we are going to present a large class of elliptic degenerated second order differential operators in R^n to which we are going to associate local operators satisfying the conditions from the preceding section.

Let X be an open set in R^n and $a_{ij} \in C^{2+\gamma}(X)$, $i,j=1,\ldots,n, \gamma>0$, $b_k \in C^2(X)$, $k=1,\ldots,n$. Assume that for each $x \in X$ and $\xi \in R^n$, $\sum_{ij} a_{ij}(x)\xi^i\xi^j \geq 0$ and the matrix $(a_{ij}(x))$ is not nul. This last condition is equivalent with the apparently stronger condition $\sum_i a_{ii}(x)>0$.

Now we state a particular case of Theorem 1.8.2 from O.A.Oleinik and E.B. Radkevich [32]. We denote by $C_{(2)}(U)$ the class of those real functions on $U \subset R^n$, which possess bounded second order derivatives.

3.1. Theorem. Let Ω be a relatively compact open subset of X such that the boundary $\partial\Omega$ is a smooth surface of class C^4 and suppose that for each point $x \in \partial\Omega$ the following relation holds

$$a_{ij}(x)n^i n^j > 0$$

where n is the normal vector to $\partial\Omega$ in x. Then there exists a constant $M>0$ such that for each $\lambda \geq M$ and each $f \in C_{(2)}(\Omega)$ the equation

$$\sum_{ij} a_{ij}\partial^2 u/\partial x_i \partial x_j + \sum_k b_k \partial u/\partial x_k - \lambda u = f$$

has a unique solution $u \in C_{(2)}(\Omega) \cap C_o(\Omega)$.

The usual maximum principle (see for example Theorem 1.1.2 from [32]) shows that the solutions from the preceding theorem satisfy the inequality

$$\sup_{\Omega} |u| \leq (\text{const.})(\sup_{\Omega} |f|).$$

The following theorem was proved by J.P. Roth in [34] IV.4.4 considering more general barrier functions.

3.2. Theorem. Let T be a locally compact space and A a linear operator $A:D(A) \longrightarrow C_0(T)$, $D(A) \subset C_0(T)$, $\overline{D(A)} = C_0(T)$ such that the following conditions are fulfilled:

(i) $(\forall)$ U open $\subset$ T, $(\forall)$ $f \in D(A)$,

$(f=0 \text{ on } U) \Longrightarrow (Af=0 \text{ on } U)$

(ii) $(\forall)$ $f \in D(A)$, $(\forall)$ $\varphi \in C_0^\infty(R)$, $\varphi(0)=0 \Longrightarrow \varphi \circ f \in D(A)$

(iii) A has a closure $\overline{A}$ which is the infinitesimal generator of a sub-Markov semigroup on $C_0(T)$.

Let Ω be a relatively compact open sent which is K_σ and such that the following conditions are fulfilled:

(iv) $(\forall)$ $x \in \partial\Omega$, $(\exists)$ U_x an open set and $\varphi_x \in D(A)$ such that $x \in U_x$, $\varphi_x(x)=0$, $\varphi_x(y)>0$ $(\forall)$ $y \in U_x \cap \overline{\Omega}$, $y \neq x$ and $A\varphi_x \leq 0$ on $U_x \cap \Omega$

(v) $(\exists)$ ϕ, $\theta \in D(A)$, ϕ, $\theta>0$, $A\phi<0$, $A\theta>0$ on $\overline{\Omega}$.

Then for each $f \in C(\partial\Omega)$ there exist a sequence $\{g_n\} \subset D(A)$ and $g \in C(\overline{\Omega})$ such that $g_n \longrightarrow g$, $A_{g_n} \longrightarrow 0$ uniformly on each compact subset of Ω and $g_{|\partial\Omega} = f$. The function g is uniquely determined by f.

Now we define the local operator L on X by putting for each open set $U \subset X$:

$$D(U,L) = \{f \in C_{(2)}(U) / Lf \in C(U)\}$$

$$Lf = \sum_{ij} a_{ij} \partial^2/\partial x_i \partial x_j f + \sum_k b_k \partial/\partial x_k f \quad .$$

L is locally dissipative and the condition from Corollary 1.9 is fulfilled. We denote by $\tilde{L}$ the "local closure" of L.

3.3. Theorem. $\tilde{L}$ has a base of P- and D- regular sets.

Proof. Let us consider an open ball U, $\overline{U} \subset X$ and a function $\varphi \in C^\infty(X)$ such that $\varphi=0$ on U and $\varphi>0$ on $X \setminus \overline{U}$. If U_1 is another open ball such that $\overline{U}_1 \subset X$, $\overline{U} \subset U_1$, then the open set U_1 and the operator

$$L^1 = L + \varphi \sum_k \partial^2/\partial x_k^2$$

satisfy the requirements from 3.1. Therefore if α is large, for each $f \in C_{(2)}(U_1)$ there exists $u \in C_{(2)}(U_1) \cap C_o(U_1)$ such that $L^1_\alpha u = f$. Further if $f \in C(\bar{U}_1)$ we choose a sequence $\{f_n\} \subset C^\infty(X)$ such that $f_n \longrightarrow f$ uniform on $\bar{U}_1$. The sequence $\{u_n\} \subset C_{(2)}(U_1) \cap C_o(U_1)$ defined by $L^1_\alpha u_n = f_n$ is a Cauchy sequence in $C_o(U_1)$. The limit $u = \lim u_n$ satisfies $u \in D(U_1, \tilde{L}^1) \cap C_o(U_1)$ and $\tilde{L}^1_\alpha u = f$. Thus U_1 is P-regular for $\tilde{L}^1_\alpha$ on account of Proposition 1.10 and of the relation $C_o^\infty(U_1) \subset D(U_1, L^1)$. Then from 1.12.1° we deduce, via the Hille-Yosida theorem, that $A = L^1_\alpha$, $D(A) = C_o^\infty(U_1)$ fulfil conditions (i), (ii), (iii) from Theorem 3.2. Let $x_o \in U$ and $\rho > 0$ such that $\{\|x - x_o\| < \rho\} \subset U$. The function $v(x) = \rho^2 - \|x - x_o\|^2$ satisfy $L^1_\alpha v(x) = -2\sum_i a_{ii} - 2\sum_k (x^k - x_o^k) b_k - \alpha v(x)$ on U. Then there exists $\varepsilon > 0$ such that for each $\rho < \varepsilon$ and each $x_o \in U$ such that $\{\|x - x_o\| < \varepsilon\} \subset U$ the function v satisfy $L^1_\alpha v < 0$ on $\{v > 0\} = \{\|x - x_o\| < \rho\}$. Then for a fixed $\Omega = \{\|x - x_o\| < r\}$ with $r < \varepsilon$ and $\bar{\Omega} \subset U$ we can check condition (iv) from 3.2 i.e.: for each $x \in \partial\Omega$ we put $g_x(y) = \rho^2 - \|y - x_o + (\rho - r)(x - x_o)\|^2$ where ρ is choosen such that $r < \rho < \varepsilon$ and $\{g_x > 0\} \subset U$ and define $\varphi_x = g_x$ in a suitable neighbourhood of x. Condition (v) from 3.2 can be checked using functions ψ and θ which are defined by $\psi(y) = \rho^2 - \|y - x_o\|^2$ and $\theta(y) = \|y - x_o\|^2 + \sigma$ on Ω. In order that $L\theta > 0$ on $\bar{\Omega}$ we should impose a new condition on the diameter of Ω and take r and σ small enough. We conclude from Theorem 3.2 that small enough open balls in U are D-regular for $\tilde{L}^1_\alpha$. From 1.12.6° we get that such balls are also P-regular for $\tilde{L}^1_\alpha$. But $L_1 = L$ on U, and hence these balls are P-regular for $\tilde{L}_\alpha$. To complete the proof we apply 1.12.2°, 3° and 5°.

Remark. If instead of Theorem 3.1 from above one uses the Theorem of J.P. Roth from [35], which has a more straight forward proof, one can get a result similar to Theorem 3.3 but for another class of differential operators.

Now we present a result which allows us to transform a local operator preserving the property that it possesses a base of P-regular sets.

3.4. Proposition

Let L be a locally dissipative local operator on a locally compact space, X, which has a countable base of P-regular sets. If a, $c \in C(X)$, $a>0$, $c\geq 0$, then the local operator, L^o, defined by

$$D(U,L_o)=D(U,L) \text{ for each open set } U$$

$$L^o f=aLf-cf \text{ for each } f\in D(U,L),$$

is also locally dissipative. If U is P-regular for L then it is also P-regular for L^o.

Proof. L^o is obviously locally dissipative. First we suppose $c\equiv 0$. Then G^{oU}, the kernel associated to L^o on U, is given by

$$G^{oU}f=G^U(f/a) \text{ for each } f\in C_b(U) \;.$$

Now we assume $c>0$. Then $L^o=c((a/c)L-1)$ and the assertion results from the first case and $1.12.1^o$.

In the general case, when $c\geq 0$, we firstly deduce that U is P regular with respect to L_λ for each $\lambda>0$ and from $1.12.3^o$ we get the required conclusion.

4. Excessive functions

Let X be a locally compact space with a countable base, L a locally dissipative local operator on X which has a base of P- and D-regular sets and $(\Omega, M, M_t, X_t, \theta_t, P^x)$ the standard process constructed in Theorem 2.5.

4.1. Proposition. If U is D-regular for L, then each point $x\in \partial U$ is regular, i.e. $E^x[T_{CU}>0]=0$, and

$$H^U f(x)=P_{CU}f(x) \quad \text{for each } x\in \overline{U},\ f\in C(X) \;. \tag{1}$$

Proof. If there exists a P-regular set, V, such that $\overline{U}\subset V$, then relation (1) is a consequence of relation 2.1, (2). Therefore we have a base of D-regular sets which fulfill (1). Let U be an arbitrary D-regular set and let $\{U_n/n\in N\}$ be a countable covering of U of D-regular sets which fulfil (1) and $\overline{U}_n\subset U$ for any $n\in N$. We also suppose that for each $k\in N$ the set $\{n\in N/U_n=U_k\}$ is infinite. Further we put $T_k=T_{CU_k}$ and $R_o=0$, $R_{k+1}=R_k+T_{k+1}\circ\theta_{R_k}$. The sequence $\{R_k\}$ being increasing, $\{X_{R_k}\}$ converges a.s. on $\{\sup_k R_k<\xi\}$. Let us suppose that

$X_{R_k}(\omega) \longrightarrow x \in U$ and let n_o be such that $x \in U_{n_o}$. For any large n we have $X_{R_n}(\omega) \in U_{n_o}$. On the other hand $X_{R_{n+1}} = X_{T_{n+1}} \circ \theta_{R_n}$ and $X_{T_{n+1}} \in CU_{n_o}$, a.s., provided $U_{n_o} = U_{n+1}$. Therefore we conclude $\lim_{k\to\infty} X_{R_k} \in CU$, a.s. on $\{\sup R_k < \xi\}$. Further we deduce $\lim_{k\to\infty} R_k = T_{CU}, P^x$ - a.s. for any $x \in U$.

Now let $f \in C(\partial U)$ and $h = H^U f \in C(\bar{U})$. Since for $y \in U_{n+1}$ we have $h(y) = H^{U_{n+1}} h(y)$ we deduce for each $n \in N$,

$$E^x[h(X_{R_n})] = E^x[E^{X_{R_n}}[h(X_{T_{n+1}})]] = E^x[h(X_{R_{n+1}})] .$$

Then $h(x) = E^x[h(X_{R_n})] = \lim_{k\to\infty} E^x[h(X_{R_k})] = E^x[f(X_{T_{CU}})]$, which proves (1) for any $x \in U$. In order to prove that a point $x_o \in \partial U$ is regular we are choosing a D-regular set, V, that fulfils (1) and such that $x_o \in V$. Then for $f \in C_c(V)$ we put

$$\varphi(x) = E^x[\int_0^{T_{CV}} f(X_t)\,dt] \;;$$

this means $\varphi(x) = G^V f(x)$ if $x \in V$ and $\varphi(x) = 0$ if $x \in E \setminus V$. Let us denote by

$$h(x) = \begin{cases} H^U \varphi(x) & \text{if } x \in U \\ \varphi(x) & \text{if } x \in CU \end{cases} ,$$

$$g(x) = E^x[\varphi(X_{T_{CU}})], \qquad x \in X.$$

We have $h \in C_c(E)$ and $h(x) = g(x)$ except possible for those points $x \in E \setminus U$ which are not regular for $E \setminus U$, i.e. except for a semipolar set [6] p.80. Therefore for each $\lambda > 0$ we have

$$E^x[\int_0^\infty \exp(-\lambda t) h(X_t)\,dt] = E^x[\int_0^\infty \exp(-\lambda t) g(X_t)\,dt] \quad ,$$

and hence $h(x) = \lim_{\lambda\to\infty} \lambda E^x[\int_0^\infty \exp(-\lambda t) g(X_t)\,dt]$ for each $x \in X$.

Let us suppose that x_o is not regular, i.e. $E^{x_o}[T_{CU} > 0] = 1$. Then we have

$$\lim_{t\to 0} E^{x_o}[g(X_t)] = \lim_{t\to 0} E^{x_o}[g(X_t);\; t < T_{CU}].$$

On the other hand, the strong Markov property allows us to deduce

$$E^{x_o}[g(X_t);\; t < T_{CU}] = E^{x_o}[\varphi(X_{T_{CU}});\; t < T_{CU}] .$$

When $t \longrightarrow 0$ the last term tends to $g(x_o)$, and hence $g(x_o)=$ $=\lim_{t\to 0} E^{x_o}[g(x_t)]$. Finally $g(x_o)=\lim_{\lambda\to\infty} E^{x_o}[\int_0^\infty \exp(-\lambda t)g(X_t)dt]$. The function f being arbitrary this shows that x_o is in fact regular.

The following result was proved by the author in [39]. It is related to a conjecture of T.Watanabe [41, II].

4.2. Theorem. Let $(\Omega, M, M_t, Y_t, \theta_t, P^x)$ be a standard process with state space E and $\mathcal{U}$ a base of open sets. If a universally measurable function $s:E \longrightarrow R_+$ satisfies

(a) $P_{CU}s \leq s$ for each $U \in \mathcal{U}$

(b) $s(x)=\lim P_{CU}s(x)$ for each x, where the limit is taken over the directed set $\{U \in \mathcal{U}/x \in U\}$ when $U \searrow \{x\}$,

then s is excessive.

Proof. We consider a metric d on E and for each fixed $n \in N$, $n \geq 1$, choose a sequence $\{D_i/i \in N\}$ of open sets and another sequence $\{U_i/i \in N\} \subset \mathcal{U}$ such that

$$\bigcup_{i\in N} D_i=E,\quad \overline{D}_i \subset U_i\ ,\quad d(U_i)<1/n\ ,\quad (\forall)\quad i \in N$$

and the set $\{i \in N/U_i \cap K \neq \emptyset\}$ is finite for any compact set K. We define $R(\omega)=T_{CU_i}(\omega)$ if $X_o(\omega) \in D_i \setminus \bigcup_{j=1}^{i-1} D_j$, then put $R_o=0$, $R_1=R$ and $R_{k+1}=$ $=R_k+R\circ\theta_{R_k}$ for each $k \in N$, $k \geq 1$.

$\{R_k/k \in N\}$ are stopping times and $\lim_{k\to\infty} R_k=\xi$. The function $s_n : E \longrightarrow \overline{R}_+$, defined by $s_n(x)=\inf \{P_{CU_i}s(x)/i \in N,\ x \in U_i\}$ is universally measurable. Further let $x_o \in E$, $t>0$, $n \in N$, $n \geq 1$. We are going to prove the following inequality by induction:

(2) $$s(x_o) \geq E^{x_o}[s_n(X_t);\ t \leq R_k]+E^{x_o}[s(X_{R_k});\ R_k<t]\ .$$

For k=0 it is trivial. Further (a) implies:

(3) $$s(x) \geq E^x[s(X_R)]\ .$$

On the other hand we have

$$E^x[s(X_{T_{CU_i}}); t<T_{CU_i}]=E^x[P_{CU_i}s(X_t); t<T_{CU_i}]\geq$$

$$\geq E^x[s_n(X_t); t<T_{CU_i}].$$

and hence $E^x[s(X_R); t-r<R]\geq E^x[s_n(X_{t-r}); t-r<R]$.

In this inequality we put $x=X_{R_k}(\omega)$ and $r=R_k(\omega)$ and integrate over $\{\omega/R_k(\omega)<t\}$ with respect to $dP^{x_o}(\omega)$:

(4) $$\int\chi_{\{\omega/R_k(\omega)<t\}}\int s(X_R(\omega'))\cdot\chi_{\{\omega'/t-R_k(\omega)<R(\omega')\}}dP^{X_{R_k}(\omega)}(\omega')dP^{x_o}(\omega)$$

$$\geq\int\chi_{\{\omega/R_k(\omega)<t\}}\int s_n(X_{t-R_k(\omega)}(\omega'))\chi_{\{\omega'/t-R_k(\omega)<R(\omega')\}}dP^{X_{R_k}(\omega)}(\omega')dP^{x_o}(\omega).$$

Using the strong Markov property[1], we can rewrite the last term as

$$E^{x_o}[s_n(X_t); t-R_k<R\circ\theta_{R_k}; R_k<t]$$

Now in (3) we put $X_{R_k}(\omega)$ instead of x and integrate both sides of (3) over $\{\omega/R_k(\omega)<t\}$:

$$E^{x_o}[s(X_{R_k}); R_k<t]\geq$$

$$\geq\int\chi_{\{\omega/R_k(\omega)<t\}}\int s(X_R(\omega')dP^{X_{R_x}(\omega)}(\omega')dP^{x_o}(\omega),$$

further, using (4) we get

$$\geq E^{x_o}[s_n(X_t); R_k<t<R_{k+1}]+$$

$$+\int\chi_{\{\omega/R_k(\omega)<t\}}\int s(X_R(\omega'))\chi_{\{\omega'/R(\omega')\leq t-R_k(\omega)\}}dP^{X_{R_k}(\omega)}(\omega')dP^{x_o}(\omega).$$

Again the strong Markov property shows that this last term equals

$$E^{x_o}[s(X_{R_{k+1}}); R_{k+1}\leq t].$$

---

1) We have used the strong Markov property in the following form: if τ is a stopping time and $G(\omega,\omega')$ an $M_\tau\times F$ measurable non-negative function, then

$$E^x[G(..\theta_\tau(.)/M_\tau](\omega)=E^{X_\tau(\omega)}[G(\omega,.)].$$

Thus we have

$$(5)\qquad E^{x_o}[s(X_{R_k});\ R_k<t]\geq$$

$$E^{x_o}[s_n(X_t);\ R_k<t<R_{k+1}]+E^{x_o}[s(X_{R_{k+1}});\ R_{k+1}\leq t]\ .$$

Now let us suppose that (2) is valid; from (2) and (5) we get

$$s(x_o)\geq E^{x_o}[s_n(X_t);\ t<R_{k+1}]+E^{x_o}[s(X_t);\ t=R_{k+1}]+$$

$$+E^{x_o}[s(X_{R_{k+1}});\ R_{k+1}<t\ ],$$

which leads to formula (2) with k+1 instead of k.

Letting $k\to\infty$ we have

$$s(x_o)\geq E^{x_o}[s_n(X_t)]\ .$$

Since condition (b) implies $s=\lim_{n\to\infty} s_n$, we obtain

$$s(x_o)\geq\liminf_{n\to\infty} E^{x_o}[s_n(X_t)]\geq E^{x_o}[s(X_t)]\ .$$

If $U\in\mathcal{U}, x_o\in\mathring{U}$, then

$$s(x_o)\geq\limsup_{t\to 0} E^{x_o}[s(X_t)]\geq\liminf E^{x_o}[s(X_t)]\geq$$

$$\lim_{t\to 0} E^{x_o}[P_{CU}s(X_t);\ t<T_{CU}]=P_{CU}s(x_o)\ ,$$

and hence $s(x_o)=\lim_{t\to 0} E^{x_o}[s(X_t)]$.

From this theorem we deduce the following characterisation for the excessive functions.

4.3. Corollary. A universally measurable function $s:X\to R_+$ is excessive if and only inf,

(7) $H^U s\leq s$ on U, for each D-regular set, U,

(8) $s(x)=\sup\,\{H^U s(x)\,|\,U$ is D-regular, $x\in U\}$ for each $x\in X$.

Proof. If s is an excessive function, relation (7) results from (1) and Proposition (2.8) from [6] p.73. On the other hand if $\{U_n\}$ is a sequence of D-regular neigbourhoods of a point x such that their diameters converges to 0 (in an arbitrary metric compatible with the topology), then $X_{T_{CU_n}} \to x$, P^x-a.s. Then from Proposition (2.12), [6] p.75 we know that $s(X_{T_{CU_n}}) \to s(x)$, P^x - a.s., and hence $s(x) \leq \liminf H^{U_n}s(x)$, which leads to (8). Conversely let us suppose that s satisfies (7) and (8). If U and V are D-regular and $\bar{U} \subset V$, then from (7) we deduce $H^V s \leq H^U s$ on U. Therefore for $x \in X$,

$$\sup \{H^U s(x) \mid U \text{ is D-regular, } x \in U\} = \lim_{U \searrow \{x\}} H^U s(x)$$

and we apply Theorem 4.2.

Now we give a property of "piecing out" for excessive functions. It is proved here in order to show that axiom (H_3) considered in chapter II is a necessary condition for the construction of the structure presented in that chapter.

4.4. Proposition. Let s, s' be two excessive functions and U a D-regular set. If $v(x) = \inf (s(x), H^U s(x) + s'(x) - H^U s'(x))$ for $x \in U$ and $v(x) = s(x)$ if $x \in CU$, then v is an excessive function.

Proof. We put $H = H^U$ and $T = T_{CU}$. If $x \in CU$ we have $v(x) \geq E^x[v(X_t)]$ for any $t > 0$. If $x \in U$ and $t > 0$ are given we have:

$$s'(x) \geq E^x[s'(X_t); t < T] + E^x[s'(X_T); t \geq T],$$

$$Hs'(x) = E^x[s'(X_T)] = E^x[E^{X_t}[s'(X_T)]; t < T] + E^x[s'(X_T); t \geq T].$$

Further we get

$$Hs(x) + s'(x) - Hs'(x) \geq E^x[v(X_t); t < T] + E^x[s(X_T); t \geq T].$$

The last term on the right hand side may be written

$$\int \chi_{\{t \geq T(\omega)\}} v(X_T(\omega)) dP^x(\omega) \geq$$

$$\geq \int\int \chi_{\{t \geq T(\omega)\}} v(X_{t-T(\omega)}(\omega')) dP^{X_T(\omega)}(\omega') dP^x(\omega) =$$

$$= \int \chi_{\{t \geq T(\omega)\}} v(X_{t-T(\omega)}(\theta_T(\omega))) dP^x(\omega) = E^x[v(X_t); t \geq T].$$

Hence we have $v(x) \geq E^x[v(X_t)]$. The relation $\lim_{t \to 0} E^x[v(X_t)] = v(x)$ is a consequence of the regularity of U.

II. QUASIHARMONIC SPACES

1. Definitions

Let X be a locally compact space with a countable base and $\mathcal{H}=\{\mathcal{H}(U)/U$ open set in $X\}$ a family of vector spaces of real continuous functions which defines a sheaf on X. A relatively compact open set, U, such that $\partial U\neq\Phi$ will be called regular if the following conditions are satisfied:

1° for any $f\in C(\partial U)$ there exists a unique function $u\in C(\overline{U})$ such that

$$u_{|\partial U}=f \quad \text{and} \quad u_{|U}\in\mathcal{H}(U) ,$$

2° when the function f is positive the associated function u is positive too.

For a regular set U, we define a kernel on X, denoted by H^U, in the following way: if $f\in C(X)$ then

$$H^U f(x)=\begin{cases} f(x) & \text{if } x\in X\setminus U \\ u(x) & \text{if } x\in U \end{cases}$$

where $u\in C(\overline{U})$, $u_{|\partial U}=f_{|\partial U}$ and $u_{|U}\ \mathcal{H}(U)$. If D is an open set such that $\overline{U}\subset D$, then there is an evident way to restrict H^U to D and we preserve the same notation for the restricted kernel. Let D be an open set and $f\in\mathcal{B}_b(D)$; f is called supermedian if:

$$H^U f\leq f , \quad (\forall)\ U \text{ regular}, \qquad \overline{U}\subset D.$$

A supermedian function f on D is called superharmonic if:

$$f(x)=\sup\ \{H^U f(x)/U \text{ regular},\ \overline{U}\subset D,\ x\in U\}, \qquad (\forall)\quad x\in D.$$

(It should be noted that the above definition restrict our attention to bounded superharmonic functions, but this notion is fairly good for the purposes of this paper). The set of all superharmonic functions on D is denoted by $\mathcal{S}_b(D)$. An element $s\in\mathcal{S}_b(D)$ is called strict if:

$$H^U s(x)<s(x), \quad (\forall)\quad U \text{ regular}, \quad \overline{U}\subset D, \quad (\forall)\quad x\in U,$$

We remark that the limit of a bounded monotone sequence of supermedian functions is supermedian too; besides the limit is superharmonic provided that the sequence is increasing and its elements are superharmonic.

1.1. Definition

The pair $(X,\mathcal{H})$ is called a quasiharmonic space if the following axioms are fulfilled:

(H_1) There exists a base of regular sets.

(H_2) If D is an open set and U a regular set, $\overline{U}\subset D$ and $s,t\in S_b(D)$ then

$$\inf\,(s,\ H^U s+t-H^U t)\in S_b(D)\ .$$

(H_3) If D is an open set and $t\in B_b(D)$ and if $\mathcal{U}$ is a base of regular sets of D such that

1° $\overline{U}\subset D$, $H^U t\leq t$, $(\forall)$ $U\in\mathcal{U}$

2° $t(x)=\sup\{H^U t(x)/U\in\mathcal{U},\ x\in U\}$, $(\forall)$ $x\in D$, then $t\in S_b(D)$.

(H_4) $1\in S_b(X)$.

(H_5) $(\forall)$ $x\in X$, $(\exists)$ V an open set, $x\in V$ and a strict element $s\in S_b(V)\cap C(V)$.

From now on we suppose that $(X,\mathcal{H})$ is a quasiharmonic space. We remark that $F(x)$, the family of all regular neighbourhoods of the point x may be viewed as a directed set and from (H_4) we deduce that

$$\lim_{F(x)} H^{Ux}=\varepsilon x \qquad \text{vaguely, } (\forall)\ x\in X.$$

It results that $\inf(s,t)\in S_b(X)$ provided that $s,t\in S_b(X)\cap C(X)$.

If U, V are regular sets and $\overline{U}\subset V$ then $H^U H^V=H^V$; if s is supermedian then $H^V s\leq H^U s$. Further we note the following consequence of (H_3): if s_1, $s_2\in B_b(X)$ and $s_1+s_2\in S_b(X)$ and $H^U s_i\leq s_i$, $i=1,2$, $(\forall)$ $U\in\mathcal{U}$, where $\mathcal{U}$ is a base of regular sets, then s_1 , $s_2\in S_b(X)$.

Let now $f\in B_b(X)$ be such that $f\geq H^U f$, $(\forall)$ $U\in\mathcal{U}$, where $\mathcal{U}$ is a base of regular sets. We put the following notation

$$\hat{f}(x)=\sup\,\{H^U f(x)/U\in F(x)\cap\mathcal{U}\} \qquad (\forall)\quad x\in X.$$

If we consider a metric on X we can construct coverings $\mathcal{U}_n$ of X, $\mathcal{U}_n\subset\mathcal{U}$ $n\in N$ such that each $U\in\mathcal{U}_n$ has a diameter smaller than $1/n$. Putting $f_n=\inf\,\{H^U f/U\in\mathcal{U}_n\}$ we get $\hat{f}=\lim f_n$ and deduce that $\hat{f}\in B_b(X)$; then we get $\hat{f}\in S_b(X)$.

Let D be an open set and $f\in B_b(D)$. We shall say that f is balanced on an open set $A\subset D$ if $H^U f=f$, $(\forall)$ U regular, $\overline{U}\subset A$. From (H_3) we get a largest open set $G\subset D$ such that f is balanced on G. We denote by

$$\text{bsupp}\ f=D\setminus G.$$

A potential on D is an element $p\in S_{b+}(D)$ such that:

$$s\in S_b(D),\ s\leq p,\ \text{bsupp}\ s=\Phi \Longrightarrow s\leq 0.$$

The set of all potentials on D is denoted by $P_b(D)$. For a function $f\in B_b(X)$, we shall use the notation:

$$Rf=\inf\{s\in S_{b+}(X)/s\geq f\}.$$

The specific order on $S_{b+}(D)$ (D open set) is denoted by "$\preccurlyeq$":

$$s,\ t\ S_{b+}(D),\quad s\preccurlyeq t \text{ iff } t-s\in S_{b+}(D);$$

Now we introduce some notations that will be used in proofs. First we denote by F_0 a fixed countable base of regular sets; then for each $n\in N$ we denote by $\mathfrak{S}_n$ the family of all functions $\sigma:\{0,1,\ldots n\}\to F_0$ and put $\mathfrak{S}=\bigcup_n \mathfrak{S}_n$. If $\sigma\in\mathfrak{S}_n$ the restriction of σ to $\{0,1,\ldots k\}$ $(k<n)$ is denoted by σ_k (hence $\sigma_k\in\mathfrak{S}_k$); if $U\in F_0$ then σ_U is the element from $\mathfrak{S}_{n+1}$ defined by

$$\sigma_{U,n}=\sigma \qquad \text{and} \qquad \sigma_U(n+1)=U.$$

1.2. <u>Proposition</u>

1° Let $s\in S_b(X)$, then there exists a unique pair h, p $S_b(X)$ such that

$$s=h+p\ ,$$

p is a potential and h is balanced on X. This is called the Riesz decomposition of s.

2° If $\{p_n/n\in N\}\subset P_b(X)$ and $\sum_n p_n$ is bounded then $\sum_n p_n\in P_b(X)$.

<u>Proof</u>. We put

(1) $$h=\inf\{H^{\sigma(n)}H^{\sigma(n-1)}\ldots H^{\sigma(0)}s/\sigma\in\mathfrak{S}_n\ ,\quad n\in N\}.$$

From (H_4) we get $-||s||<h<s$. Then we deduce

$$H^U h=h \quad \text{for every } U\in Fo$$

and from (H_3) we deduce that h, $-h\in S_{b+}$. Further we characterise the elements t of $P_b(X)$ by menas of the following relation:

(2) $$\inf\{H^{\sigma(n)}H^{\sigma(n-1)}\ldots H^{\sigma(0)}t/\sigma\in\mathfrak{S}_n\ ,\ n\in N\}=0\ .$$

The rest of proof goes as in the case of a harmonic space.

1.3. <u>Propositon</u>. If $s\in S_{b+}(X)$, $t\in P_b(X)$ and $s-t\in S_b(X)$, then $s-t\geq 0$.

<u>Proof</u>. If s=h+p is the Riesz decomposition of s, then $s-t\geq$ $\geq\inf\{H^{\sigma(n)}H^{\sigma(n-1)}\ldots H^{\sigma(0)}(s-t)/\sigma\in\mathfrak{S}_n\ ,\ n\in N\}=h\geq 0$.

1.4. <u>Proposition</u>. Let X be noncompact and $s\in S_b(X)\cap C_0(X)$. If the cone

$$C=\{t\in S_{b+}(X)/t \text{ lower semicontinuous}\}$$

linearly separates the points of X, then $s\in\mathcal{P}_b(X)$.

Proof. Lemma 1.5 from below shows that $s\geq 0$. Let $s=h+p$ be the Riesz decomposition of s. From (1) we deduce that h is upper semicontinuous and $0\leq h\leq s$. Again by Lemma 1.5 we deduce $-h\geq 0$, and hence $h=0$.

The following lemma adapts Bauer's wellknown minimum principle to our situation (see [1]).

1.5. Lemma. Let C be a convex cone of locally lower bounded, lower semicontinuous numerical functions on X, such that:

1° $(\forall)$ $x\in X$, $(\exists)$ $s\in C$, $s>0$, $s(x)<\infty$.

2° $(\forall)$ $x, y\in X$, $x\neq y$, $(\exists)$ s, $t\in C$ and $s(x)t(y)\neq s(y)t(x)$.

Let f be a locally lower bounded, lower semicontinuous numerical function on X such that

$$\liminf_{x\to\infty} f(x)\geq 0 \qquad \text{(if X is not compact).}$$

We suppose that for each $x\in X$ there exists μ_x a positive Radon measure such that

1° $\mu_x(1)\leq 1$

2° $\mu_x(s)\leq s(x)$, $(\forall)$ $s\in C$

3° $f(x)\geq\mu(f)$

4° supp $\mu\neq\{x\}$.

Then $f\geq 0$.

Proof. Let $\alpha=\inf f$ and we suppose that $\alpha<0$. Then we put $T=\{x\in X/f(x)=\alpha\}$; T is a compact set. We denote by

$$S=\{s_{|T}-\beta/s\in C \quad , \beta\in R_+\}$$

If x_o is a point in the Choquet boundary ∂T (it is well known that $\partial T\neq\emptyset$; see for example Lemma 2 in [13] p.26 or II.2.7 in [7]) then $\mu_{x_o}(1)\leq 1$ and $f(x_o)=\alpha\geq\mu_{x_o}(f)\geq\alpha$. Hence $\mu_{x_o}(T)=1$ and we deduce that $\mu_{x_o}=\varepsilon_{x_o}$, which contradicts 4°.

2. Construction of balayages and of potential kernels

Let U be a regular set and $f\in\mathcal{B}_b(X)$. We put

$$M^U f=\sup\,(f, H^U f).$$

This is Mokobodzki's regularisation operator, used in [30] p.213. If $\sigma\in\mathfrak{S}_n$ we put

$$M_\sigma f = M^{\sigma(n)} M^{\sigma(n-1)} \dots M^{\sigma(0)} f.$$

We remark that $\{M_\sigma f / \sigma \in \mathcal{G}\}$ is upper directed and countable and put

$$Mf = \sup \{M_\sigma f / \sigma \in \mathcal{G}\}.$$

2.1. Proposition. Let $f \in \mathcal{B}_b(X)$ be such that

(1) $$f(x) \leq \limsup_{U \in F(x) \cap F0} H^U f(x) , \qquad (\forall) \quad x \in X.$$

Then $Mf \in S_b(X)$. If $f \geq 0$ then $Mf = Rf$, and bsupp $Rf \subset$ bsupp f.

Proof. If $U \in F0$ and $\sigma \in \mathcal{G}$ then $M^U M_\sigma f = M_{\sigma_U} f$ and we get $M^U Mf = Mf$, and hence $Mf \geq H^U Mf$.

Let $x \in X$. If $f(x) = Mf(x)$, then (1) implies

(1') $$MF(x) = \sup \{H^U Mf(x) / U \in F(x) \cap F0\}.$$

If $Mf(x) > f(x)$, for a given $\varepsilon > 0$, we choose $n \in N$ and $\sigma \in \mathcal{G}_n$ such that

$$M_\sigma f(x) > Mf(x) - \varepsilon \quad \text{and} \quad M_\sigma f(x) > f(x).$$

Let $p = \inf \{k < n / M_{\sigma_k} f(x) = M_\sigma f(x)\}$; then

$$M_\sigma f(x) = H^{\sigma(p)} (M_{\sigma_{p-1}} f)(x) < H^{\sigma(p)} Mf(x),$$

where the case $p=0$ is considered under the convention $M_{\sigma_{-1}} f = f$. Again (1') is checked. Using (H_3) we deduce $Mf \in S_b(X)$. On the other hand, if $s \in S_b(X)$ then $M^U s = s$ for each $U \in F0$ which leads to $Ms = s$; it results $MF = Rf$. If $H^U f = f$ then $H^U Rf \in S_b(X)$ (use (H_2)) and $H^U Rf > f$. Hence $H^U Rf = Rf$, which finishes the proof.

2.2. Theorem. Let $s, t \in S_{b+}(X)$, then $R(s-t) = M(s - \inf (s,t)) \in S_{b+}(X)$ and $R(s-t) \not\prec s$.

Proof. We have $R(s-t) = R(s - \inf (s,t))$ and 2.1 shows that

$$R(s-t) = M(s - \inf (s,t)) \in S_{b+}(X).$$

Let now $r = R(s-t)$ and U be a regular set, then

$$s - t \leq r, \quad s - t \leq H^U r + s - H^U s,$$

and hence $s - t \leq \inf (r, H^U r + s - H^U s)$;

(H_2) shows that $r \leq H^U r + s - H^U s$ or $H^U(s-r) \leq s-r$. On the other hand, for each $x \in X$ we have

$$\sup \{H^U(s-r)(x)/U \in F(x)\} = \lim_{U \in F(x)} H^U(s-r)(x) = s(x) - r(x),$$

which shows $s-r \in S_{b+}(X)$.

Now we are going to introduce the balayage on open sets: let A be an open set and $s \in S_{b+}(X)$; we denote by

$$R^A s = \inf \{t \in S_{b+}(X)/t \geq s \text{ on } A\}$$

2.3. Proposition. Let A be an open set and $s,t \in S_{b+}(X)$. Then

1° $R^A s = M(\chi_A . s) \in S_{b+}(X)$

2° $R^A(s+t) = R^A s + R^A t$.

Proof

1° is obvious. For 2° we follow Mokobodzki [30] I.14. The inequality $R^A(s+t) \leq R^A s + R^A t$ is obvious. Then we put $u = R(R^A(s+t) - R^A s)$, $v = R^A(s+t) - u$ and from 2.2 we know that $u, v \in S_{b+}(X)$. One easily deduces $R^A(s+t) = u+v$, $u \leq R^A t$, $v \leq R^A s$, and hence $v = R^A s$, $u = R^A t$ on A, which implies $v = R^A s$, $u = R^A t$.

Property 2° from this proposition allows us to define the balayage B^A over the set A as a linear operator on $S_{b+}(X) - S_{b+}(X)$: if $s,t \in S_{b+}(X)$ and $f = s-t$ then we define

$$B^A f = R^A s - R^A t$$

2.4. Proposition

Let K be a closed set, A be an open set and $f,g \in S_{b+}(X) - S_{b+}(X)$. If $K \subset A$ and $f \geq g$ on $A \setminus K$ then $B^A f \geq B^A g$ on $X \setminus K$.

Proof. We suppose $f,g \in S_{b+}(X)$ and put $X' = X \setminus K$ and for $s \in S_{b+}(X')$ denote by $R's = \inf \{t \in S_{b+}(X')/t \geq s \text{ on } X' \cap A\}$. It is easy to see that $(R^A s)_{|X'} = R'(s_{|X'})$ for any $s \in S_{b+}(X)$. From 2.3, 1° there results $R'(f_{|X'}) \geq R'(f_{|X'})$.

2.5. Proposition. Let $f \in P_b(X) - P_b(X)$ and A be an open set such that bsupp $f \subset A$. Then $B^A f = f$.

Proof. Let $s,t \in P_b(X)$ be such that $f = s-t$. If U is a regular set and $\overline{U} \cap \text{bsupp } f = \emptyset$ or $\overline{U} \subset A$ then

$$R^A t + f \geq H^U(R^A t + f).$$

From 1.3 we get $R^A t+f \in \mathcal{P}_b(X)$, and hence

$$R^A t+f \geq R^A s.$$

Repeating this argument for $-f$ we get the converse inequality: $R^A s-f \geq R^A t$. Now we begin the construction of the kernel associated to a given potential. The construction makes use of the fixed family $\mathcal{F}_o$, but the kernel does not depend on the particular choice of $\mathcal{F}_o$ (see (3) below). For each open set D we denote by

$$\mathfrak{S}_n(D)=\{\sigma \in \mathfrak{S}_n / \overline{\sigma(k)} \subset D,\ k=0,1,\dots n\},$$

$$\mathfrak{S}(D)=\bigcup_{n\in N} \mathfrak{S}_n(D).$$

If $s\in S_{b+}(X)$ and $\sigma \in \mathfrak{S}_n(D)$ we define

(2) $$P_\sigma s=M(H^{\sigma(n)}P_{\sigma_{n-1}}(s)+s-H^{\sigma(n)}s)$$

$$P_{\sigma_0} s=M(s-H^{\sigma(o)}s)$$

Relation (2) can be written $P_\sigma s=M(s-H^{\sigma(n)}(s-P_{\sigma_{n-1}}s))$, and hence $P_\sigma s\in S_{b+}(X)$, $P_\sigma s \preccurlyeq s$. Thus $P_{\sigma_{k-1}}s \leq P_{\sigma_k}s$ for $k\leq n$. We deduce that the family $\{P_\sigma s/\sigma \in \mathfrak{S}(D)\}$ is upper directed and denote by $P(D)s$ its supremum. Further we deduce $P(D)s\in S_{b+}(X)$, $P(D)s \preccurlyeq s$, bsupp $P_\sigma s\subset D$ and bsupp $P(D)s\subset D$.

Let now $U\in \mathcal{F}_o$, $\overline{U}\subset D$. Then

$$P_\sigma s \leq H^U P_\sigma s+s-H^U s \leq P_{\sigma_U} s,$$

and hence $P(D)s \leq H^U P(D)s+s-H^U s \leq P(D)s$,

$$\text{or} \quad s-P(D)s=H^U(s-P(D)s),$$

U being arbitrary we deduce bsupp $(s-P(D)s)\subset CD$.

If $s=h+p$ is the Riesz decomposition of s then we have $P_\sigma s=P_\sigma p \leq p$. Hence $P(D)s=P(D)p\leq p$ and we get $P_\sigma s$, $P(D)s\in \mathcal{P}_b$. Then we remark that $P(D)s=s$ provided $s\in \mathcal{P}_b(X)$ and bsupp $s\subset D$. Now we are going to prove the following relation

(3) $$P(D)s=\sup\{t\in \mathcal{P}_b(X)/\text{bsupp } t\subset D \text{ and } t\preccurlyeq s\}.$$

First we remark that (2) shows $P_\sigma t \leq P_\sigma s$, provided that $t\in S_{b+}(X)$ and $t\preccurlyeq s$. Hence $P(D)t<P(D)s$. If in addition we suppose that $t\in \mathcal{P}_b(X)$ and bsupp $t\subset D$, there results

$t=P(D)t\leq P(D)s$, which leads to (3).

The next sentence shows that the sup in (3) may be taken in the specific order: if $t\in S_{b+}(X)$ and $t\preccurlyeq s$, then $P(D)t\preccurlyeq P(D)s$. To prove this sentence we remark that

$$P_\sigma t\preccurlyeq t\preccurlyeq s, \quad (\forall)\ \sigma\in\Theta(D), \text{ and hence}$$

$$P(D)s-P_\sigma t\geq H^U(P(D)s-P_\sigma t),$$

provided that U is regular and $\overline{U}\cap \text{bsupp } P_\sigma t=\Phi$ or $\overline{U}\subset D$. From (H_3) and 1.3 we deduce $P(D)s-P_\sigma t\in P_b(X)$, which leads to the desired conclusion.

If D_1, D_2 are two open sets such that $D_1\subset D_2$, then $P(D_1)s=P(D_1)P(D_2)s\preccurlyeq P(D_2)s$. If $\{D_n/n\in N\}$ is an increasing sequence of open sets, then $P(D_n)s$ increases to $P(\bigcup_n D_n)s$. The following proposition is very useful in proving the construction theorem.

2.6. <u>Proposition</u>. Let $s\in S_{b+}(X)$ and t_1, $t_2\in P_b(X)$, t_1, $t_2\preccurlyeq s$ bsupp $t_1\cap$bsupp $t_2=\emptyset$. Then $t_1+t_2\preccurlyeq s$.

<u>Proof</u>. We choose two open sets D_1, D_2 such that $D_1\cap D_2=\emptyset$ and bsupp $t_i\subset D_i$, $i=1,2$. Then $t_i=P(D_i)t_i\preccurlyeq P(D_i)s$. On the other hand $P(D_1)s=P(D_1)(s-P(D_2)s)\preccurlyeq s-P(D_2)s$, which finishes the proof.

2.7. <u>Theorem</u>. If $q\in P_b(X)$ then there exists a kernel V on X such that $V1=q$, $V(A)\in P_b(X)$ and bsupp $(V(A))\subset\overline{A}$, for each Borel set A.

<u>Proof</u>. Let A, B be open sets. We are going to prove

$$P(A)q+P(B)q=P(A\cup B)q+P(A\cap B)q.$$

Let t_1, $t_2\in P_b(X)$ be such that bsupp $t_1\subset A$, bsupp $t_2\subset B$ and t_1, $t_2\preccurlyeq q$. Then we choose two open sets, D and G, such that $D\cap G=\emptyset$ and

$$\text{bsupp } t_1\subset D, \qquad \text{bsupp } t_2\setminus A\subset G.$$

We get

$$\text{bsupp } P(G)t_2\cap\text{bsupp } t_1=\Phi \text{ and bsupp } (t_2-P(G)t_2)\subset A\cap B;$$

further

$$P(G)t_2+t_1\preccurlyeq P(A\cup B)q \text{ and}$$

$$t_2-P(G)t_2\preccurlyeq P(A\cap B)q. \text{ Hence } t_1+t_2\leq P(A\cup B)q+P(A\cap B)q.$$

From (3) there results

$$P(A)q+P(B)q\leq P(A\cup B)q+P(A\cap B)q.$$

Now we consider t_1 , $t_2\in \mathcal{P}_b(X)$, such that t_1 , $t_2\preccurlyeq q$, bsupp $t_1\subset A\cap B$, and bsupp $t_2\subset A\cup B$. Then we choose two open sets, D and G, such that $D\cap G=\Phi$ $G\subset B$, bsupp $t_1\subset D$ and bsupp $t_2\setminus A\subset G$. There results bsupp $P(G)t_2\cap$bsupp $t_1=\Phi$ and

$$\text{bsupp }(t_2-P(G)t_2)\subset A\ ,$$

which leads to

$$P(G)t_2+t_1\preccurlyeq P(B)q \text{ and}$$

$$t_2-P(G)t_2\preccurlyeq P(A)q.$$

Hence $t_1+t_2\leq P(A)q+P(B)q$, which implies

$$P(A\cup B)q+P(A\cap B)q\leq P(A)q+P(B)q$$

on account of (3).

Now we put $V(A)=P(A)q$ for each open set A and from [27] III.25 deduce that V defines a kernel on X. The monotone class theorem applied to the family

$$\{A \text{ Borel set}/V(A)\in \mathcal{P}_b(X),\quad V(X\setminus A)\in \mathcal{P}_b(X)\}$$

shows that it is identical to the family of all Borel sets. The rest of proof is obvious.

2.8. Theorem. Let V_1 , V_2 be two kernels on X such that $V_1(X)=V_2(X)=q\in\mathcal{P}_b(X)$ and $V_i(A)\in\mathcal{P}_b(X)$, bsupp $V_i(A)\subset\bar{A}$ for any Borel set A and i=1,2. Then $V_1=V_2$.

Proof. We adapt Meyer's arguments from [26]. Let K_1 , K_2 be compact sets such that $K_1\cap K_2=\Phi$; then we choose $\sigma\in\tilde{\mathcal{G}}_n$ such that either $\overline{\sigma(i)}\cap K_1=\Phi$ or $\overline{\sigma(i)}\cap K_2=\Phi$, for any i=0,1,...n. We are going to prove

(4) $\quad H^{\sigma(n)}..H^{\sigma(0)}q+V_1(CK_1)-V_2(K_2)=H^{\sigma(n)}...H^{\sigma(0)}q+V_2(CK_2)-V_1(K_1)\geq 0.$

To do we suppose it is true for σ_{n-1} , and $\overline{\sigma(n)}\cap K_2=\Phi$. Then we get

$$H^{\sigma(n)}H^{\sigma(n-1)}...H^{\sigma(0)}q+H^{\sigma(n)}V_1(CK_1)-H^{\sigma(n)}V_2(K_2)\geq 0\ ,$$

which leads to (4) because $H^{\sigma(n)}V_2(K_2)=V_2(K_2)$ and $V_1(CK_1)\geq H^{\sigma(n)}V_1(CK_1)$ From (4) we get $V_1(CK_1)\geq V_2(K_2)$ and $V_2(CK_2)\geq V_1(K_1)$ on account of 1.(2), which leads to $V_1=V_2$.

2.9. Proposition. If $q \in P_b(X) \cap C(X)$ then the kernel, V, constructed in Proposition 2.7 verifies $VC_b(X) \subset C_b(X)$.

Proof. If f is lower semicontinuous then $H^U f$ and Mf are alike. Hence $P_\sigma q$ and P(D)q are lower semicontinuous. Then Vf is lower semicontinuous provided f is nonnegative, bounded and lower semicontinuous. If $0 \leq f \leq 1$, $f \in C(X)$ then V(f) and V(1-f) are lower semicontinuous and their sum is continuous, thus teyselves are continuous.

2.10. Proposition. Let $q \in P_b(X) \cap C(X)$ and let V be its associated kernel. If $s \in S_{b+}(X)$ is lower semicontinuous, then s is a V-dominant i.e.: if $f,g \in B_+(X)$ and $Vf \leq Vg+s$ on $\{f>0\}$ then $Vf \leq Vg+s$ on X.

Proof. We use standard arguments (see [27] X.T.4):

$Vf=\sup\{V\varphi/\varphi$ upper semicontinuous, $0 \leq \varphi \leq f$, $\{\varphi>0\}$ is compact$\}$ and $Vg=\inf\{V\varphi/\varphi$ lower semicontinuous, $g \leq \varphi\}$.

Thus we may suppose that $K=\{f>0\}$ is compact, f is upper semicontinuous and bounded, and g is lower semicontinuous. In this case s+Vg-Vf is lower semicontinuous and for each $\varepsilon>0$ we have

$$s+Vg-Vf+\varepsilon>0$$

on an open set A, with $K \subset A$. On account of 2.4, 2.5 one deduces

$$s+Vg+\varepsilon \geq R^A(s+Vg+\varepsilon) \geq R^A Vf=Vf$$

2.11. Corollary

Let $q \in P_b(X) \cap C(X)$ and let V be the kernel associated to q. There exists a unique sub-Markov resolvent $\{V_\lambda/\lambda \geq 0\}$ on the space $C_b(X)$ such that $V_0=V$. Moreover if $s \in S_{b+}(X)$ is lower semicontinuous then $\lambda V_\lambda s \leq s$, for each $\lambda>0$.

Proof. The resolvent is given by Theorem I.1.11. If we put $f=(s-\lambda V_\lambda s)^+$ and $g=(s-\lambda V_\lambda s)^-$, then the inequality $\lambda V_\lambda s \leq s$ becomes $\lambda Vf \leq \lambda Vg+s$ and now it follows easily from Proposition 2.10.

2.11'. Notation

Let $p \in P_b(D)$, D open set $\subset X$ and let V be the kernel on D, associated to p. If $f \in B_b(D)$ then we denote by

$$f \cdot p=Vf.$$

2.12. Remark

Let p_1, $p_2 \in P_b(X)$ and D be an open set such that

bsupp $(p_1-p_2)\subset CD$. If $f\in\mathcal{B}_b(X)$ is such that $\{f\neq 0\}\subset D$, then $f\cdot p_1=f\cdot p_2$. To see this we have only to prove the case $f=\chi_A$, with A opent set $\subset D$, and this is obvious from the construction of the two kernels.

2.13. <u>Proposition</u>. Let $p\in\mathcal{P}_b(X)$ and U be a regular set such that $p-H^Up\in\mathcal{P}_b(U)$. Then

$$f\cdot p-H^U(f\cdot p)=f_{|U}\cdot(p-H^Up) \text{ on } U, \qquad (\forall)\quad f\in\mathcal{B}_b(X).$$

<u>Proof</u>. We denote by V the kernel on U, defined by

$$V(A)=\chi_A\cdot p-H^U(\chi_A\cdot p).$$

Then $p-H^Up-V(D)$ is balanced on U and positive. Hence $p-H^Up=V(U)$. From 2.8 we deduce

$$V(A)=\chi_A\cdot(p-H^Up), \qquad (\forall)\quad A \text{ Borel set}, \quad A\subset U.$$

Now let A be a Borel set $A\subset X$ and $A_1=A\cap U$, $A_2=A\setminus U$. Since $\chi_{A_2}\cdot p-H^U(\chi_{A_2}\cdot p)\leq p-H^Up$ and $\chi_{A_2}\cdot p-H^U(\chi_{A_2}\cdot p)$ is balanced we get it is nul. Then $\chi_A\cdot p-H^U(\chi_A\cdot p)=\chi_{A_1}\cdot(p-H^Up)$.

3. <u>An inequality</u>. Here we prove an inequality that will be very useful in the rest of this paper. This inequality is analogous to one ot R.M. Blumenthal and R.K. Getoor (see pages 240-241 of [6]).

3.1. <u>Proposition</u>. Let K be a closed set and A an open set and let $g=p_1-p_2$, p_1 , $p_2\in S_{b+}(X)$. Then for each $f\in\mathcal{P}_b(X)-\mathcal{P}_b(X)$ such that bsupp $f\subset K$ the following inequality is satisfied

$$B^Af(x)\leq(\sup_{\Gamma}|f-B^Af|)p_2(x), \qquad (\forall)\quad x\in X$$

provided that $\Gamma=V\setminus K$ and V is an open set such that $K\subset V$ and either $g\geq 1$ on V and $g\leq 0$ on A or $g\geq 1$ on A and $g\leq 0$ on V.

<u>Proof</u>. We may consider $\overline{V}\cap\overline{A}=\phi$ without loss of generality. Let us suppose that $g\geq 1$ on V and $g\leq 0$ on A. If $f\in\mathcal{P}_b(X)-\mathcal{P}_b(X)$, bsupp $f\subset K$, then 2.5 gives us $B^Vf=f$ and 2.4 shows that $B^Vf=B^VB^\Gamma f$ on $X\setminus K$.

On the other hand 2.3, 1° shows that $B^VB^\Gamma=B^\Gamma$ hence $f=B^\Gamma f$ on A. Then

$$B^Af=B^AB^\Gamma f=B^AB^\Gamma(f-B^Af)+(B^AB^\Gamma)B^Af=\;\dots\;,$$

$$B^Af=\sum_{n=1}^{\infty}(B^AB^\Gamma)^n(f-B^Af), \tag{1}$$

because $(B^AB^\Gamma)^nf\to 0$ when $n\to\infty$.

But $f-B^Af \leq (\sup_\Gamma |f-B^Af|)q$ on Γ, hence $B^Af \leq (\sup_\Gamma |f-B^Af|) \sum_{n=1}^{\infty} (B^AB^\Gamma)^n q$.

Since $B^Aq \leq 0$, for $n \geq 1$ we get

$$(B^AB^\Gamma)^n q \leq (B^AB^\Gamma)^{n-1} B^A (B^\Gamma q - q) \leq (B^AB^\Gamma)^{n-1} B^A (p_2 - B^\Gamma p_2)$$

$$\leq (B^AB^\Gamma)^{n-1} p_2 - (B^AB^\Gamma)^n p_2 \quad ,$$

which leads to the desired inequality.

Now we suppose that $q \geq 1$ on A and $q \leq 0$ on V. Then $(B^AB^\Gamma)^n 1 \leq (B^AB^\Gamma)^{n-1} B^A q \leq (B^AB^\Gamma)^{n-1} (B^A q - q)$ because $B^\Gamma q \leq 0$; furhter we get

$$(B^AB^\Gamma)^n 1 \leq (B^AB^\Gamma)^{n-1} (p_2 - B^A p_2) \quad .$$

On the other hand

$$(B^AB^\Gamma)^n (f-B^Af) \leq (\sup_\Gamma |f-B^Af|) (B^AB^\Gamma)^n 1$$

and from (1) we obtain the desired inequality, because

$$\sum_{n=1}^{m} (B^AB)^{n-1} (p_2 - B^A p_2) = p_2 - \sum_{n=0}^{m-1} (B^AB^\Gamma)^n B^A (p_2 - B^\Gamma p_2) - (B^AB^\Gamma)^m p_2 \leq p_2 .$$

For a standard process the inequality takes the following form.

3.1'. <u>Theorem</u>. Let $(\Omega, \mathcal{M}, \mathcal{M}_t, X_t, \theta_t, P^x)$ be a standard process with state space E. Let A, B be nearly Borel sets and assume that there exist two finite excessive functions P_1 , P_2 such that either $p_1 - p_2 \leq 0$ on A and $p_1 - p_2 \geq 1$ on B or $p_1 - p_2 \geq 1$ on A and $p_1 - p_2 \leq 0$ on B. Then for each $x \in E$ there exists a positive measure μ_x such that $\mu_x(E \setminus B^{-f}) = 0$, where B^{-f} is the fine closure of B,

$$\mu_x(1) \leq p_2(x) \text{ and } P_A f(x) = \mu_x(f - P_A f),$$

for each $f = s_1 - s_2$, with s_1 , s_2 natural potentials such that $P_B f = f$. If the process has continuous paths, then the measures can be choosen such that $\mu_x(E \setminus \partial B) = 0$ for each $\in$ E.

<u>Proof</u>. First we define inductively a sequence of stopping times: $R_0 = 0$, $R_{2k+1} = R_{2k} + T_A \circ \theta_{R_{2k}}$, $R_{2k+2} = R_{2k+1} + T_B \circ \theta_{R_{2k+1}}$. Let us suppose that $p_1 - p_2 \leq 0$ on A and $p_1 - p_2 \geq 1$ on B. Then $(p_1 - p_2)(X_{R_{2k}}) \geq 1$ and $(p_1 - p_2(X_{R_{2k+1}}) \leq 0$. Therefore $P^x(R_{2k} < \infty) \leq E^x [(p_1 - p_2)(X_{R_{2k}}) - (p_1 - p_2)(X_{R_{2k-1}})]$.

Since $\{p_1(X_t)\}$, $\{p_2(X_t)\}$ are supermartingales we deduce $P^x(R_{2k}<\infty)\leq$ $\leq E^x[p_1(X_{R_{2k}})-p_1(X_{R_{2k-1}})]$ and $\sum_{k=1}^{\infty} P^x(R_{2k}<\infty)\leq \sum_{k=1}^{\infty} E^x[p_1(X_{R_{2k}})-p_1(X_{R_{2k-1}})]\leq p_1(x)$. Therefore $R_{2k}\to\infty$, a.s.

Now if s is a natural potential we have $(P_AP_B)^k s(x)=E^x[s(x_{R_{2k}})]\to$ $\to 0$, and hence

$$P_Af=\sum_{n=1}^{\infty}(P_AP_B)^n(f-P_Af) \ .$$

We define the measure $\mu_x(\cdot)=\sum_{n=1}^{\infty}(P_AP_B)^n(\cdot)(x)$ and the remainder proof follows as in 3.1. Another form of this inequality is given in the following proposition.

3.2. Proposition. Let K be a compact set and U a regular one, $K\subset U$ and p_1, $p_2\in S_{b+}(X)$, $g=p_1-p_2$ such that $g\geq 1$ on an open set V, $K\subset V$ and $g=0$ on CU. Then

$$H^Uf(x)\leq(\sup_{\Gamma}|f-H^Uf|)p_2(x), \qquad (\forall)\quad x\in X,$$

where $\Gamma=V\setminus K$ and $f\in P_b(X)-P_b(X)$, sbuspp $f\subset K$. The proof of this proposition is similar to the proof of the first case in Proposition 3.1 (put H^U in the place of B^A).

The functions p_1, p_2 from the above propositions may be obtained in a standard way: Let $p\in P_b(X)\cap C(X)$ be a strict superharmonic function. The space

$$T=\{f=s-t/s,t\in P_b(X)\cap C(X),\ \text{supp } f \text{ is compact}\}$$

is a vector lattice, and $p-H^Up\in T$, for any regular set U, hence T linearly separates the points in X. Then for each pair (K,A) such that K is compact, A is open and $\overline{A}\cap K=\emptyset$, there exists p_1, $p_2\in P(X)\cap C(X)$ such that $g=p_1-p_2\geq 1$ on K and $g=0$ on A.

4. C-strict potentials. The notion of a C-strict potential considered here was introduced in [13] p.166 under the name "strict potential". We recall this defintion in an equivalent form:

4.1. Defintion. A potential $p\in P_b(X)\cap C(X)$ is called C-strict potential if: for each pair (μ,υ) of positive Radon measures such that

a) $\mu(p)=\upsilon(p)<\infty$,

b) $\mu(s)\leq\upsilon(s)$, $\quad(\forall)\quad s\in P_{b+}(X)\cap C(X)$, there results $\mu=\nu$.

It is easy to see that any C-strict potential is a strict superharmonic function. If there exists $p \in P_b(X) \cap C(X)$ a strict superharmonic function, then using standard arguments one may construct a C-strict potential. Namely one chooses a sequence $\{f_n=s_n-t_n / n \in N\} \subset T$ (T is the space from the final part of the preceding section) such that:

($\forall$) V open set, ($\forall$) $f \in C_c(V)$, ($\forall$) $\varepsilon>0$ there exists $n \in N$ such that $f_n \in C_c(V)$ and $|f_n-f|<\varepsilon$.

Then $q=\sum_{n\in N} (1/2^n)(1/\|s_n\|+\|t_n\|)(s_n+t_n)$ is a C-strict potential (In fact q is hyperstrict in the meaning of [13] p.167).

4.2. Proposition

Let p be a C-strict potential and $\{V_\lambda/\lambda\geq 0\}$ its associated resolvent (i.e. $V_0 f=f\cdot p$, see 2.11). If $s \in S_{b+}(X)$ is lower semicontinuous then s is excessive: $\lambda V_\lambda s\leq s$ and $s=\lim_{\lambda\to\infty} \lambda V_\lambda s$.

Proof

We follow C.Constantinescu and A.Cornea [13] p.250. The inequality $\lambda V_\lambda s\leq s$ is known from 2.11. Further we suppose that $s \in C(X)$. If $\alpha<\beta$ we have $\alpha V_\alpha s-\beta V_\beta s=(\alpha-\beta)V_\beta(s-\alpha V_\alpha s)\leq 0$, thus the limit $\lim_{\alpha\to\infty} \alpha V_\alpha s(x)$ exists for each point $x \in X$.

Now let x be fixed and define $\varphi(f)=\lim_{\alpha\to\infty} \alpha V_\alpha f(x)$ on the space $F=\{f=t_1-t_2/t_i \in P_b(X) \cap C(X),\ i=1,2\}$. φ is a positive functional and $\varphi(1)\leq 1$, therefore φ is also a bounded functional when we provide F with the uniform norm. Since $C_c(X)\subset \bar{F}$, the extension of φ to $\bar{F}$ defines a measure, μ, on X such that $\varphi(f)=\mu(f)$ for each $f\in C_c(X)\cap F$.

From relation (2) 1.2, for $s\in P_b(X)\cap C(X)$ we get a sequence $\{s_n\} \subset P_b(X) \cap C(X)$ such that $s_n\searrow 0$, $s_n\leq s$ and $s-s_n\in C_c(X)$. Then $\varphi(s)=\varphi(s_n)+\mu(s-s_n) \longrightarrow \mu(s)$, because $\varphi(s_n)\leq s_n(x)\longrightarrow 0$. Hence $\mu(s)=\varphi(s)\leq s(x)$. On the other hand $\varphi(p)=\lim_{\alpha\to\infty} \alpha V_\alpha V1(x)=\lim_{\alpha\to\infty} (V1(x)-V_\alpha 1(x))=$ $=p(x)$. The measures μ and ε_x satisfy a) and b) from 4.1, which implies $\mu=\varepsilon_x$. We conclude $\alpha V_\alpha s(x)\longrightarrow s(x)$ for each $s \in P_b(X)\cap C(X)$. Further we deduce $\alpha V_\alpha \varphi(x)\longrightarrow\varphi(x)$ for each $\varphi\in C_c(X)$. Assume now that $s \in S_b+(X)$ is lower semicontinuous. For $\varepsilon>0$ we choose $\varphi \in C_c(X)$ such that $\varphi\leq s$ and

$s(x)-\varepsilon\leq\varphi(x)$. Then we get

$$s(x)-\varepsilon\leq\lim_{\alpha\to\infty}\alpha V_\alpha\varphi(x)\leq\lim_{\alpha\to\infty}\alpha V_\alpha s(x)\leq s(x),$$

which leads to $\lim_{\alpha\to\infty}\alpha V_\alpha s(x)=s(x)$.

4.3. Theorem

Let p be a C-strict potential, $s\in P_b(X)\cap C(X)$ and K be a closed set such that $K\cap\text{bsupp } s=\emptyset$. Suppose further that one of the following conditions is fulfilled:

1° K is compact and $\lim_{x\to\infty} s(x)=0$ (when X is not compact).

2° bsupp s is compact.

Then there exists a sequence $\{\varphi_n/n\in N\}\subset C_{b+}(X)$ such that $\text{supp }\varphi_n\cap K=\emptyset$, $\varphi_n.p\leq s$ and $\lim_{n\to\infty}\varphi_n.p=s$ in $C_b(X)$.

Proof

Assume first that 1° is fulfilled. Let $\{V_\lambda/\lambda\geq 0\}$ be the resolvent associated to p. Then we have

$$\alpha V_\alpha s\nearrow s \quad\text{in}\quad C_o(X).$$

Since $\alpha V_\alpha s=\alpha V(s-\alpha V_\alpha s)$, putting $f_n=n(s-nV_n s)$, we get $s=\lim_{n\to\infty} f_n.p$.

Let now A, B be two open sets such that $\bar{A}\cap\bar{B}=\emptyset$, $K\subset B$, bsupp $s\subset A$ and $X\setminus A$ is compact. From 3.1 we get a constant C such that

$$B^A f\leq C||f-B^A f||,\qquad(\forall)\quad f\in P_b(X)-P_b(X)\quad\text{such that bsupp } f\subset\bar{B}.$$

Particularly we get $||g.p||\leq(1+C)||g.p-B^A(g.p)||$, $(\forall)$ $g\in C_c(B)$.
We choose $g\in C(X)$ such that $\text{supp } g\subset B$, $0\leq g\leq 1$ and $g=1$ on an open neighbourhood of K.
Then $||(gf_n).p||\leq(1+C)||(gf_n).p-B^A((gf_n).p)||\leq(1+C)||f_n.p-B^A(f_n.p)||$.
On the other hand for each $\varepsilon>0$, there is $n_o\in N$ such that $f_n.p+\varepsilon\geq s$, for $n\geq n_o$; thus

$$B^A(f_n.p)+\varepsilon\geq B^A s=s\quad\text{and}$$

$$0\leq f_n.p-B^A(f_n.p)\leq s-B^A(f_n.p)\leq\varepsilon\ .$$

Then we have $||(g.f_n).p||\longrightarrow 0$. Hence the sequence $\{\varphi_n=(1-g)f_n/n\in N\}$ satisfies the conclusions of the theorem. The case when 2° is fulfilled is simillar.

4.4. Proposition. Let $t\in P_b(X)$ and U be a regular set such that $p=t-H^U t$ is a C-strict potential on U. For $s\in P_b(U)\cap C(U)$ such that bsupp s is compact, there exists $s'\in P_n(X)$ such that

$$\text{bsupp } s'=\text{bsupp } s \text{ and } s=(s'-H^U s')_{|U} \quad .$$

If t is lower semicontinuous then s' is also lower semicontinuous. If $t\in C(X)$ then $s'\in C(X)$. If X is noncompact and $\lim_{x\to\infty} t(x)=0$ then $\lim_{x\to\infty} s'(x)=0$.

Proof. Let K be a closed set such that $\partial U\subset \overset{\circ}{K}$ and bsupp $s\cap K=\emptyset$. On the space U we aply 4.3, 2° and get a sequence $\{\varphi_n/n\in N\}\subset C_{c+}(U)$ such that supp $\varphi_n\cap K=\emptyset$, $\varphi_n.p\leq s$ and $\lim_{n\to\infty}\varphi_n.p=s$ in $C_b(U)$.

On the other hand, from 3.2 we get a constant C such that for any $f\in P_b(X)-P_b(X)$,

$$H^U f\leq C||f-H^U f||, \quad \text{provided that bsupp } f\subset U\setminus\overset{\circ}{K} \quad .$$

Thus we deduce

$$||\varphi.t||<(C+1)||\varphi.t-H^U(\varphi.t)||=(C+1)||\varphi.p|| \quad ,$$

for any $f\in C_c(U)$ such that supp $\varphi\cap\overset{\circ}{K}=\emptyset$.

Then $\varphi_n.t$ is a Cauchy sequence in $B_b(X)$. We put

$$s'=\lim_{n\to\infty}\varphi_n.t$$

From 2.13 we have $\varphi_n.p=(\varphi_n.t-H^U(\varphi_n.t))_{|U}$, which gives us $s=(s'-H^U s')_{|U}$ and the proof is finished.

4.5. Proposition. A function $p\in P_b(X)\cap C(X)$ is a C-strict potential if and only if:

$$(\forall)\quad U \text{ open set,}\quad (\forall)\quad f\in C_c(U),\qquad (\forall)\quad \varepsilon>0,$$

$$(\exists)\quad \varphi\in C_c(U) \text{ such that } ||f-\varphi.p||<\varepsilon \quad .$$

Proof. Suppose p is C-strict and let $f\in C_c(U)$, $\varepsilon>0$. We choose $p', p''\in P_b(X)\cap C(X)$ such that

$$||f-p'+p''||<\varepsilon/2 \quad \text{and} \quad p',p'' \in H(X\setminus \text{supp } f) .$$

From 4.3 we get φ_1 , $\varphi_2 \in C_{c+}(U)$ such that

$$||p'-\varphi_1 . p||<\varepsilon/4 \ , \quad ||p''-\varphi_2 . p||<\varepsilon/4$$

Then we put $\varphi=\varphi_1-\varphi_2$ and obtain $||f-\varphi . p||<\varepsilon$. The converse is easier.

4.6. Proposition

1° If p is a C-strict potential and U a regular set, then $p-H^U p$ is a C-strict potential on U.

2° Let $p \in P_b(X)$ and let $\mathcal{U}$ be a covering of regular sets such that $p-H^U p$ is a C-strict potential on U, for any $U \in \mathcal{U}$. Then

$(\forall)$ V open set, $(\forall)$ $f \in C_c(V)$, $(\forall)$ $\varepsilon>0$,

$(\exists)$ $\varphi \in C_c(V)$ such that $||f-\varphi . p||<\varepsilon$.

Proof

1° Is an easy consequence of 4.5 and 2.13.

2° First we consider the case when U is an open set such that $\overline{U} \subset W$, for suitable $W \in \mathcal{U}$. Let $f \in C_c(U)$; then we choose a sequence $\{\varphi_n / n \in N\} \subset C_c(U)$ such that $|f-\varphi_n . q| \longrightarrow 0$ on W, where $q=p-H^W p$ is viewed as a C-strict potential on W. Then we have

$$|f-\varphi_n . p| \leq |f-H^W(\varphi_n . p)|+|H^W(\varphi_n . p)| \quad \text{on } X.$$

From 3.2 we get $|H^W(\varphi_n . p)| \leq C(\sup_{V \setminus U} |\varphi_n . q|)$, where V is an open set such that $\overline{U} \subset V$, $\overline{V} \subset W$, (the function g from 3.2 may be taken of the form $g=\alpha(p-H^W p)$). Thus $||H^W(\varphi_n . p)|| \longrightarrow 0$, and hence $||\varphi_n . p-f|| \longrightarrow 0$.

The general case follows by decomposing each f in a sum of functions such that their supports lie in open sets of the type considered in the above case.

5. Construction of functions with locally given singularities

This section is not related to the reminder.

Let U be a regular set, and $p \in P_b(X) \cap C(X)$ be such that $q=p-H^U p$ is a C-strict potential on U and suppose K is a compact set, $K \subset U$. We denote by

$$H(K)=\{f \in C(X) \ / \ (\exists) \quad t \in P_b(X), \quad |f| \leq t, \quad f \in H(X\setminus K)\}$$

$$H(K,U)=\{f\in C_o(U)/f\in \mathcal{H}(U\setminus K)\}$$

Obviously H(K,U) endowed with the uniform norm is a Banach space. In order to show that H(K) is also a Banach space we need the following proposition.

5.1. Proposition

If $f\in H(K)$, $s\in P_b(X)$ and $s\geq||f||$ on an open set, V, such that $K\subset V$, then $s\geq|f|$ on X.

Proof. Let $t\in P_b(X)$ be such that $|f|\leq t$. One may suppose that $||t||=||f||$ (consider $t'=\inf(t, ||f||)$; then t' is supermedian and $\hat{t}'\in P_b(X)$, $|f|<\hat{t}'$). Assume $t_{|X\setminus K}=h+q$ is the Riesz decomposition on the space $X\setminus K$. Then we define $h':X\longrightarrow R_+$ in the following way:

$$h'(x)=\begin{cases} h(x) & \text{if} \quad x\in X\setminus K \\ t(x) & \text{if} \quad x\in K \end{cases}$$

Then h' is supermedian and $\hat{h}'\in P_b(X)$, $\hat{h}'\geq|f|$, $||f||>\hat{h}'$ on X and bsupp $h'\subset K$. Thus we have $s\geq R^V\hat{h}'=\hat{h}'\geq|f|$ on X (on account of 2.5).

5.2. Proposition

H(K) is a Banach space.

Proof. Let $\{f_n/n\in N\}$ be a Cauchy sequence. Then $\lim_{n\to\infty} f_n=f\in C(X)$. Further we suppose that $||f_{n+1}-f_n||<1/2^n$. On the other hand since $p>0$ on K there is $\alpha>0$ such that $p\geq\alpha$ on an open set V such that $K\subset V$. Then from 5.1 we get $|f_{n+1}-f_n|\leq(1/\alpha)(1/2^n)p$; if $t\in P_b(X)$ and $|f_1|\leq t$ we get $|f|\leq t+(1/\alpha)p$, and hence $f\in H(K)$.

Now we state the main theorem.

5.3. Theorem

The operator $T:H(K)\longrightarrow H(K,U)$ defined by $Tf=f-H^Uf$ is a Banach space homeomorphism.

Proof. It is not difficult to see that T is injective and continuous. Thus, on account of the closed graph theorem, we have only to prove it is surjective.

First let $f\in H(K,U)$, $f\geq0$. We are looking for a function $g\in H(K)$ such that $Tg=f$. If $\psi\in C_{c+}(U)$ is such that $\psi=1$ on K and

$\{W_k/k=1\ldots n\}$ is a finite covering of K of regular sets such that $\overline{W}_k \subset U$ for any k, then we put

$$q_1 = H^{W_n} H^{W_{n-1}} \ldots H^{W_1}(\psi.q)$$

and choose $\beta>0$ which satisfies $\beta(\psi.q-q_1)\geq f$ on an open set V such $K\subset V$. On U we define the function

$$f' = \inf(\beta\psi.q-f,\ \beta q_1).$$

Then $f' \in \mathcal{C}_0(U) \cap \mathcal{S}_b(U)$, because $f'=\beta q_1$ on V and $\beta\psi.q-f,\ \beta q_1 \in \mathcal{S}_b(U\setminus K)$. From 1.4 we get $f'\geq 0$. On $U \setminus \bigcup_{k=1}^{n} W_k$ we have $f'=\beta\psi.q-f$, and hence $f' \in \mathcal{H}(U\setminus M)$, where $M=\operatorname{supp}\psi \cup (\bigcup_{k=1}^{n}\overline{W}_k)$. From 4.4 we get a potential $g' \in \mathcal{P}_b(X) \cap \mathcal{C}(X)$ such that $Tg'=f'$ and $g'=\mathcal{H}(X\setminus M)$. Now we denote by

$$g''(x) = \begin{cases} f'(x)-(\beta\psi.q)(x)-f(x) & \text{if } x \in U \\ 0 \quad \text{if } x \in X\setminus U \end{cases}$$

Then $g'' \in \mathcal{C}_c(X)$, supp $g'' \subset U$ and

$$f = g''-f'+\beta\psi.q \quad \text{on } U$$

We put $g=g''-g'+\beta\psi.p$ and it is not difficult to see that $g \in H(K)$ and and $Tg=f$.

Now we consider the case when f is not positive. We choose $\varphi \in \mathcal{C}_{c+}(U)$ such that $\varphi=1$ on K; then $\varphi.q>0$ on K and we choose $\alpha>0$ such that $\alpha\varphi.q\geq\|f\|$ on an open set V and $K\subset V$. From 5.1 we have $\alpha\varphi.q\geq f$. But $T(\alpha\varphi.p)=\alpha\varphi.q$ (see 2.13) and $f=f_1+\alpha\varphi.q$, where $f_1 \in H(K_1,U)$, $K_1=\operatorname{supp}\varphi$ and f_1 is positive. The proof is finished.

Now we remark that a careful inspection of the above construction, can give us an evaluation for $\|T^{-1}\|$. Particularly we need not the use of the closed graph theorem.

6. Construction of local operators on a quasiharmonic space

In order to construct a nice local operator associated to our quasiharmonic space we need the following result.

6.1. Proposition

Let V be an open sent, $p \in \mathcal{P}_b(X)$, and $\{V_i/i\in I\}$ a covering of V such that, for any $i \in I$, V_i is regular, $\overline{V}_i \subset V$, $p-H^{V_i}p \in C_o(V_i)$, and $p-H^{V_i}p$ is a C-strict potential on V_i. Assume that $\{s_i/i \in I\}$ is a family such that $s_i \in C(V_i) \cap P_b(V_i)$ and $s_i - s_j \in H(V_i \cap V_j)$ for any i, $j \in I$ and let W be a relatively compact open set such that $\overline{W} \subset V$. Then there exists $t \in P_b(X)$ such that bsupp $(t-s_i) \cap W \cap V_i = \emptyset$ for any $i \in I$. If p is lower semicontinuous then t is lower semicontinuous; if $p \in C(X)$ then $\in C(X)$; if X is noncompact and $\lim_{x\to\infty} p(x)=0$ then $\lim_{n\to\infty} t(x)=0$.

<u>Proof</u>. Let $\{V_{i1},\dots,V_{in}\}$ be a finite covering of $\overline{W}$ and $\{\varphi_1,\dots,\varphi_n\}$ a family of continuous functions such that $\varphi_k \in C_c(V_{i_k})$, $0 \leq \varphi_k \leq 1$, $\sum_{k=1}^{n} \varphi_k = 1$ on $\overline{W}$. From 4.4 we get the family $\{t_k/k=1,\dots,n\} \subset P_b(X)$, such that bsupp $(t_k - \varphi_k . s_{i_k}) \cap V_{i_k} = \emptyset$ and bsupp t_k=supp φ_k.

The function $t = \sum_{k=1}^{n} t_k$ satisfies the required conditions.

6.2. <u>Theorem</u>

There exists a covering $\{U_i/i \in N\}$ of regular sets of X and a family $\{p_i/i \in N\}$ such that $p_i \in C_o(U_i)$, p_i is a C-strict potential on U_i, and $p_i - p_j \in H(U_i \cap U_j)$ for any $i,j \in N$.

<u>Proof</u>. We consider a covering $\{U_i/i \in N\}$ of X, of regular sets such that for any $i \in N$ there exists V_i an open set and $r_i \in P_b(V_i) \cap C(V_i)$ such that $\overline{U}_i \subset V_i$ and r_i is a C-strict potential. Now we are going to construct a sequence $\{q_i/i \in N\}$ which fulfils the following conditions:

1° $q_n \in P_b(V_n) \cap C(V_n)$ and q_n is C-strict ,

2° for each $n \in N$, there exists a family of open sets $\{D_i^n/i=0,1,\dots,n\}$, such that $\overline{U}_i \subset D_i^n \subset V_i$, $q_i - q_j \in H(D_i^n \cap D_j^n)$ for any $i,j \leq n$.

For n=0 we put $q_o = r_o$, $D_o^o = V_o$. Then suppose that the elements q_o , $q_1,\dots,q_n$ fulfil conditions 1°, 2°. We put $K = \overline{U}_{n+1} \cap (\bigcup_{i=0}^{n} \overline{U}_i)$ and choose a relatively compact open set B such that $K\subset B, \overline{B}\subset V_{n+1}\cap(\bigcup_{i=0}^{n} D_i^n)$. In order to construct q_{n+1} and the family $\{D_i^{n+1}/i=0,1,\ldots,n+1\}$ we use 6.1 and get a function t such that $t\in P_b(V_{n+1})\cap C(V_{n+1})$, $t-q_i\in H(B\cap D_i^n)$ for $i=0,1,\ldots,n$.

Further let B_1 be an open set and $\varphi\in C(V_{n+1})$ such that $0\leq\varphi\leq 1$, $\varphi=0$ on B_1, $K\subset B_1$, $\overline{B}_1\subset B$ and $\varphi=1$ on $V_{n+1}\setminus B$. We put $q_{n+1}=t+\varphi.r_{n+1}$. Then we choose an open set A such that $A\subset V_{n+1}$, $\overline{U}_{n+1}\setminus B_1\subset A$ and $(\bigcup_{i=1}^{n}\overline{U}_i)\cap\overline{A}=\emptyset$. Then we put $D_{n+1}^{n+1}=B_1\cup A$, $D_i^{n+1}=D_i^n\setminus\overline{A}$, $i=1,\ldots,n$.

The elements $p_i=q_i-H^{U_i}q_i$ satisfy the conclusion of the theorem.

6.3. <u>Theorem</u>

Let $\{U_i/i\in N\}$ be a family of open sets and $\{p_i/i\in N\}$ a family of functions such that

a) $\bigcup_{i\in N} U_i=X$,

b) $p_i\in P(U_i)\cap C(U_i)$, p_i is a strict element, $(\forall)$ $i\in N$,

c) $p_i-p_j\in H(U_i\cap U_j)$, $(\forall)$ $i,j\in N$.

Then there exists a local operator L on X such that:

1° If $i\in N$, $\varphi\in C_b(U_i)$ then

$\varphi.p_i\in D(U_i, L)$ and $L(\varphi.p_i)=-\varphi$

2° Ker $L=H$, i.e.: $H(U)=\{f\in D(U,L)/Lf=0\}$ for any open set U.

3° L is locally closed, i.e.: for any open set V and any sequence $\{f_n/n\in N\}\subset D(V,L)$ such that

$f_n\to f$ and $Lf_n\to\varphi$ uniformly on each compact set

the limit f satisfies $f\in D(V,L)$ and $Lf=\varphi$.

4° L is locally dissipative, i.e.: if V is an open set, $f\in D(U,L)$, $x_o\in U, f(x_o)\geq 0$ and f has a maximum in x_o, then $Lf(x_o)\leq 0$.

<u>Proof</u>. Let V be an open set. D(V,L) is defined as the family of all functions $f\in C(V)$ such that for any regular set U which satisfies $\overline{U}\subset V\cap U_i$ for some $i\in N$, there exist $h\in H(U)$ and $\varphi\in C_b(U)$ such that

$f=h+\varphi.(p_i-H^U p_i)$ on U;

for such an f we put $Lf=-\varphi$ on U. Since p_i is strict we see that the definition is consistent (by using 2.13). Furhter it is not hard to prove 1°, 2°, 3°. In order to prove 4° we consider f, φ, h, U as above. Let W be a regular set, $\overline{W}\subset U$ and assume that $f\leq f(x_o)$ on W, $x_o\in W$, $f(x_o)\geq 0$. Then

$$H^W f(x_o)\leq f(x_o)H^W 1(x_o)\leq f(x_o) \quad .$$

Let us suppose that $\varphi>0$ on W. Since $\varphi.(p_i-H^U p_i)$ is strict superharmonic we have $f(x_o)<H^W f(x_o)$, which is a contradiction. This argument actually proves 4°.

6.4. Theorem

Let L be the operator constructed by Theorem 6.3 and let L' be another local operator on X.

a) If L' fulfils properties 1° and 2° from 6.3 then L=L'.

b) If the potentials p_i , $i\in N$ are C-strict and L' fulfils 1°, 3° and 4° from 6.3 then L=L'.

Proof

a) Using 2.13 we deduce $D(U,L)\subset D(U,L')$, for any open set U and $Lf=L'f$ for any $f\in D(U,L)$. Now let V be an open set, $f\in D(V,L')$, and let U be a regular set $\overline{U}\subset V$. If we put $h=f+(L'f).(p_i-H^U p_i)$, then $h\in D(U,L')$ and $L'h=0$. Hence $h\in \mathcal{H}(U)$ and $f\in D(U,L)$.

b) On account of a) we have only to prove property 2° from 6.3. Let U be a regular set such that $\overline{U}\subset U_i$ for some $i\in N$. If $\varphi\in C_{c+}(U_i)$ then using 4.3, 2° one deduces that

$$H^U(\varphi.p_i)\in D(U,L') \text{ and } L'H^U(\varphi.p_i)=0 \quad \text{on} \quad U.$$

Further from 4.5 one deduces that $H^U f\in D(U,L')$ and $L'H^U f=0$ on U for each $f\in C_c(U_i)$.

Let now V be an open set, $\overline{U}\subset V$, and let $f\in D(V,L')$, $L'f=0$. From Proposition I.1.7 one deduces $H^U f-f=0$ on U, which proves b).

If the potential p_i is C-strict and U is a regular set such that $\overline{U}\subset U_i$, then the space

$$\{\varphi.(P_i-H^U P_i)/\varphi\in C_c(U)\}$$

is dense in $C_o(U)$. Then U is P-regular in the sense of I.1.4.

Particularly we have the following:

6.5. Corollary

If $\{U_i / i \in N\}$, $\{p_i\ , i \in N\}$ satisfy conditions a), b), c) from Theorem 6.3 and all potentials $p_i\ , i \in N$ are C-strict, then the local operator constructed in 6.3 possesses a base of P- and D-regular sets.

III. Topological Properties of Global Transition Function

1. The general case. Let X be a locally compact space with a countable base and L a locally dissipative local operator on X which possesses a base of P- and D-regular sets. We shall denote by $\{P_t/t>0\}$ the global transition function of the global process $(\Omega, M, M_t, X_t \theta_t, P^x)$, given by Theorem I.2.5 and by $\{G_\lambda/\lambda>0\}$ its resolvent, i.e. $G_\lambda = \int_0^\infty \exp(-\lambda t) P_t dt$. If V is an open set, the process on V obtained by killing the global process on CV is related to the restriction of L to V by relation (4) from Theorem I.2.5 with U an arbitrary P-regular set such that $U \subset V$. A function $f: V \to \overline{R}_+$ which is excessive for this process will be called excessive on V. The kernel of L (in the sense of sheaf theory) is the sheaf Ker L={Ker L(U)/U open set}, where Ker L(U)= $=\{f \in D(U,L)/Lf=0\}$. Using the results from Chapter I one easily proves that (X, Ker L) is a quasiharmonic space. An open set, U, is D-regular for L iff it is regular for Ker L. A Borel function, $f \in \mathcal{B}_{b+}(V)$ (V open set), is excessive on V iff f is superharmonic for Ker L, on account of Proposition I.4.3. Axioms (H_1), (H_4), (H_5) are obviously satisfied. In order to check (H_2) one uses Proposition I.4.4 and for (H_3) one uses I.4.3.

Now let $\{U_n/n \in N\}$ be a countable covering of X of P- and D-regular sets. We assume that for each $k \in N$ the set $\{n/U_n=U_k\}$ is infinite and put $T_k=T_{CU_k}$, $R_0=0$, $R_{k+1}=R_k+T_{k+1}\circ\theta_{R_k}$. As in the proof of Proposition I.4.1 one deduces $\lim R_k=\zeta$. The strong Markov property shows,

$$(1) \qquad H^{U_1}\dots H^{U_k} f(x)=E^x[f(X_{R_k})], \quad (\forall)\ x\in X, \quad (\forall)\ f\in \mathcal{B}_b(X).$$

If f $\mathcal{B}_{b+}(X)$ is an excessive function, then from relation (2) of II.2.1 one deduces that f is a natural potential for the process iff f is a potential for Ker L.

In the sequel we shall use the notation $\mathcal{H}_\lambda$=Ker L_λ , i.e. $\mathcal{H}_\lambda(U)=\{f \in D(U,L) | L_\lambda f=0\}$, $\lambda \geq 0$. Of course $(X,\mathcal{H}_\lambda)$ is a quasiharmonic space and the above discution applies by substituting the global process with the corresponding λ-subprocess. Particulary $G_\lambda f$ $(f\in \mathcal{B}_{b+}(X))$ is a potential for $\mathcal{H}_\lambda$ and from Theorem II.2.8 we deduce $g\circledcirc_\lambda(G_\lambda f)=G_\lambda gf$ for each $f\in \mathcal{B}_b(X)$, where $g\circledcirc_\lambda(G_\lambda f)$ has the meaning of the notation introduced in II.2.11', with respect to the sheaf $\mathcal{H}_\lambda$.

1.1. Theorem

1° If U is a P- and D-regular set and $f \in C_b(X)$, then

(2) $$G_\lambda f = G_\lambda^U f + H_\lambda^U G_\lambda f \quad \text{on } U \quad \text{for each} \quad \lambda > 0.$$

2° If $\lambda>0$, $f \in C_c(X)$, $\varepsilon>0$ and U is an open set such that supp $f \subset U$ then there exists $g \in C_c(U)$ such that

$$|f - G_\lambda g| < \varepsilon .$$

3° Let $\mathcal{I}$ be the cone of all bounded nonnegative lower semicontinuous functions on X. $P_t \mathcal{I} \subset \mathcal{I}$ for each $t>0$.

Proof

1° Relation (2) is a consequence of relations (4) I.2.5 and (1) I.4.1.

2° Proposition II.4.6.2^o applied to the potential $P=G_\lambda 1$ with respect to $\mathcal{H}_\lambda$ proves the assertion.

3° For $\lambda>0$ and $f \in C_{b+}(X)$ relation (1) gives us:

$$H_\lambda^{U_1} \ldots H_\lambda^{U_k}(G_\lambda f)(x) = E^x[\int_{R_k}^{\infty} \exp(-\lambda t) f(X_t) dt]$$

Therefore we have $\lim_{k\to\infty} H_\lambda^{U_1} \ldots H_\lambda^{U_k}(G_\lambda f) = 0$. Further from relation (2) we deduce

$$G_\lambda f - H_\lambda^{U_i} H_\lambda^{U_{i+1}} \ldots H_\lambda^{U_k}(G_\lambda f) = G_\lambda^{U_i} f + H_\lambda^{U_i}(G_\lambda f - H_\lambda^{U_{i+1}} H_\lambda^{U_{i+2}} \ldots H_\lambda^{U_k} G_\lambda f).$$

Arguing by induction from this relation we deduce that $\{G_\lambda f - H_\lambda^{U_1} \ldots H_\lambda^{U_k} G_\lambda f \mid k \in N\}$ is an increasing sequence from $C_c(X)$ and its limit is $G_\lambda f$. Thus we get $G_\lambda f \in \mathcal{I}$. As a consequence of 2° we have $C_c(X) \subset \overline{G_\lambda(C_c(X))}$. Then on the Banach space $\overline{G_\lambda(C_c(X))}$ we apply the Hille-Yosida formula and deduce for $t>0$, $\lim_{\alpha\to\infty} \exp(\alpha t(\alpha G_\alpha - I))f = P_t f$, for each $f \in C_c(X)$. (The limit in this formula is uniform). Therefore $P_t f \in \mathcal{I}$ for each $f \in C_{c+}(X)$, which leads to $P_t \mathcal{I} \subset \mathcal{I}$.

1.2. Corollary

The process $(\Omega, \mathcal{M}, \mathcal{M}_t, X_t, \theta_t P^x)$ is a Hunt process.

Proof. This is a consequence of the relation $C_c(X)\subset G_\lambda(C_c(X))$. Let $\{T_n\}$ be an increasing sequence of stopping times and $T=\lim_{n\to\infty} T_n$. Let us suppose that T is bounded and set $L=\lim_{n\to\infty} Y_{T_n}$. If $f=R_\lambda g$, $g\in C_{c+}(X)$, then $E^x[f(X_{T_n})]=E^x[\int_{T_n}^{\infty} \exp(-_\lambda t)g(X_t)dt]\to E^x[f(X_T)]$. The supermartingale theorem shows that $f(X_{T_n})\to f(X_T)$ P^x a.s. Further one deduces that $f(X_{T_n})\to$ $\to f(X_T)$ a.s. for each $f\in C_c(X)$. On the other hand $\lim_{n\to\infty} f(X_{T_n})=f(L)$ a.s. for $f\in C_c(X)$, which implies $X_T=L$ a.s.

Next we give some alternative characterisations for C-strict potentials by means of the probabilistic notion of fine topology. This result was suggested by an example of a strict potential which is not C-strict, which A.Cornea comunicated to us.

1.3. Proposition. Let $p\in C_b(X)$ be a potential with respect to the quasiharmonic space (X, H_o). The following assertions are equivalent:

1° p is C-strict.

2° $p(x)>R^{CK}p(x)$ provided K is a compact set and a fine neighborhood of the point $x\in E$.

3° $\chi_K p\neq 0$ provided K is a compact set whose fine interior is not empty.

Proof

$1^o\Longrightarrow 2^o$ From Hunt's balayage theorem we get $R^{CK}p(x)=E^x[p(X_{T_{CK}})]$. Putting $\mu(f)=E^x[f(X_{T_{CK}})]$ we have $\mu\neq\varepsilon_x$ and $\mu(s)\leq s(x)$ for any $s\in P_b(X)\cap C(X)$. Now it is not hard to deduce 2°.

$2^o\Longrightarrow 3^o$ Let K be a compact set and suppose $\chi_K.p=0$. We choose a sequence of open sets $\{G_n\}$ such that $\overline{G}_n\subset G_{n+1}$ and $\bigcup_n G_n=X\setminus K$. Then

$$p=\sup_n \chi_{G_n}.p=\sup_n R^{CK}(\chi_{G_n}.p)=R^{CK}p\ ,$$

which leads to 3° if we assume 2°.

$3^o\Longrightarrow 1^o$ We are going to prove that $\lambda V_\lambda s\nearrow s$ for any $s\in P_b(X)\cap C(X)$, where $\{V_\lambda/\lambda\geq 0\}$ is the resolvent associated to p ($f.p=V_o f$ for any $f\in B_b(X)$). From II.2.11 we know that $\lambda V_\lambda s\leq s$. On the other hand $V_o s=\lim_{\lambda\to\infty}\lambda V_o V_\lambda s$. If we put $t=\lim_{\lambda\to\infty}\lambda V_\lambda s$ we get a lower semicontinuous function, $t \in S_b(X)$, $t \leq s$ and $(s-t) \circ p=0$. Hence $s-t$ is finely continuous and upper semicontinuous. If $s \neq t$ then for suitable $\alpha \in R_+$, $K=\{s-t \geq \alpha\}$ is a compact set with nonempty fine interior and $\chi_K \cdot p=0$, which contradicts 3°. Therefore $s=t$.

Further let μ, υ be two measures on X such that $\mu(p)=\upsilon(p)<\infty$ and $\mu(s) \leq \mu(s)$ for each $s \in P_b(X) \cap C(X)$. Then $\mu(f \circ p)=\upsilon(f \circ p)$ for any $f \in C_{b+}(X)$. Hence $\mu(s)=\lim_{\lambda \to \infty} \mu(\lambda V_\lambda s)=\lim_{\lambda \to \infty} \mu(\lambda(s-\lambda V_\lambda s) \circ p)=\upsilon(s)$ for each $s \in P_b(X) \cap C(X)$. Therefore $\mu=\upsilon$.

2. Continuity cases

We preserve the notation from the preceding section.

We shall discuss here some cases when the resolvent $\{G_\lambda / \lambda>0\}$ or the semigroup $\{Pt/t>0\}$ maps a continuous function space into itself.

First we give a simple example which shows that generally G_λ, $\lambda>0$ may transform a continuous function into a discontinuous one. Take $X=(-1,0] \times (0,1) \cup (0,1) \times (0,2) \subset R^2$ and take L to be the local closure of $\partial^2/\partial Y^2$ (see section I.3). The function $G_\lambda 1$ does not depend on the variable x either on the set $(-1,1] \times (0,1)$ or on $(0,1) \times (0,2)$ and it has a discontinuity on the set $\{0\} \times (0,1)$.

2.1. Proposition. If for some $\lambda>0$ and $g \in C_b(X)$, with $g \geq a$, $a>0$, $G_\lambda g$ is continuous, then $G_\lambda g \in D(X,L)$ and $L_\lambda G_\lambda f=-f$ for any $f \in C_b(X)$.

Proof. From II.2.9 we know $G_\lambda f=(f/g) \circledcirc_\lambda G_\lambda g \in C_b(X)$ if $f \in C_b(X)$. Further 1.1.(2) shows $G_\lambda f \in D(X,L)$ and $L_\lambda G_\lambda f=-f$.

2.2. Theorem. Assume that X iscompact and there exists a function $u>0$, $u \in D(X,L)$ and $Lu=0$. Then $\lambda G_\lambda u=u$, $G_\lambda(C(X))=D(X,L)$, and $L_\lambda G_\lambda f=-f$ for each $f \in C(X)$, $\lambda>0$.

Proof. Let $\lambda>0$ be fixed. For a P-and D-regular set, U, we compute

$$\lambda G_\lambda u-H_\lambda^U(\lambda G_\lambda u)=\lambda G_\lambda^U u=u-H_\lambda^U u \quad .$$

Since $\lambda G_\lambda u$ is a potential with respect to H_λ we deduce $u=\lambda G\lambda u+h$, where $h \geq 0$ satisfies $H_\lambda^U h=h$ for any P- and D-regular set. Since h is upper semicontinuous the next lemma shows $h=0$.

Lemma

Assume that X is compact, f is a real upper semicontinuous function on X and there is $\lambda>0$ such that $H_\lambda^U f=f$ for any P- and D-regular set. Then $f\leq 0$.

Proof. Assume x_o is a maximum point for f. Let us suppose $h(x_o)=\alpha>0$. If U is P- and D-regular $g=H^U 1-H_\lambda^U 1$ satisfies $L_\lambda g=-\lambda H^U 1\leq 0$. Moreover if U is a small enough neighbourhood of x_o then $H^U 1>0$ on U and $g=\lambda G_\lambda^U H^U 1>0$, which leads to the contradictory relation

$$1\geq H^U 1(x_o)>H_\lambda^U 1(x_o)\geq(1/\alpha)f(x_o)=1.$$

Therefore $f\leq 0$ and the lemma is proved.

In order to finish the proof of 2.2 we apply 2.1 and deduce $G_\lambda(C(X))\subset D(X,L)$. Further if $f\in D(X,L)$ then $h=f-G_\lambda(-L_\lambda f)$ satisfies $L_\lambda h=0$ and the preceding lemma implies $h=0$.

2.3. Corollary

The conditions from Theorem 2.2 imply $P_t C(X)\subset C(X)$ and $P_t u=u$ for any $t>0$.

2.4. Theorem. Assume that there exists a strict potential, p (with respect to $\mathcal{H}_o$), such that $p\in C_b(X)$.

1° Then the resolvent $\{G_\lambda/\lambda>0\}$ can be completed, namely for any $f\in C_{c+}(X)$, $G_o f=\int_0^\infty P_t f dt$ satisfies $G_o f\in C_b(X)$. Moreover for each $f\in C_c(X)$ and $\lambda\geq 0$ we have $G_\lambda f\in D(X,L)$ and $L_\lambda G_\lambda f=-f$. If $p\in C_o(X)$, then $G_\lambda f\in C_o(X)$.

2° If $p\in D(X,L)$ and $Lp<-c$ on an open set D, $c\in R$, $c>0$, then statement 1° is still valid for any $f\in C_b(X)$ provided supp $f\subset D$.

In order to prove this theorem we need the next lemma.

Lemma. If there exists a strict potential (with respect to $\mathcal{H}_o$), p, such that $p\in C(X)$, then there exists another potential (with respect to $\mathcal{H}_o$), q, such that $q\in D(X,L)$ and $Lq<0$. If X is noncompact we can choose q such that $\lim_{x\to\infty} q(x)=0$ provided $\lim_{x\to\infty} p(x)=0$.

Proof of the Lemma. We may suppose that p is C-strict without loss of generality (see II.4.1). Then for each relatively compact open set, W, we use II.6.1 and get a potential $t \in C(X)$ such that

$$t-H^U t=G^U 1, \text{ for each P- and D-regular set } U, \overline{U}\subset W.$$

If $\varphi\in C_c(W)$ then $\varphi \odot t\in D(X,L)$ and $L(\varphi\odot t)=-\varphi$. The required function may be obtained in the form $q=\sum_n \alpha_n \varphi_n \cdot t_n$, where $\{t_n/n\in N\}$ corresponds to a sequence $\{W_n/n\in N\}$ such that $\bigcup_n W_n=X$ and $\varphi_n\in C_{c+}(W_n)$, $n\in N$ are choosen such that $\bigcup_n\{\varphi_n>0\}=X$.

Proof of Theorem 2.4. The above lemma allows us to consider 1° as a particular case of 2°, namely D is an arbitrary relatively compact open set. In order to prove 2° we consider $f\in C_{b+}(X)$, supp $f\subset D$ and put $g=-Lp$ and $s=(f/g)\odot p$. Then for each P- and D-regular set, U, we have $s-H^U s=(f/g)\odot(p-H^U p)$ (see II.2.13). On the other hand $g\odot G^U 1=G^U g$ (use II.2.8) and $G^U g=p-H^U p$, and hence $s-H^U s=f\odot G^U 1=G^U f$ (on U). We deduce $s\in D(E,L)$ and $Ls=-f$. Further $L_\lambda s=-f-\lambda s\leq L_\lambda G_\lambda f$ and using II.1.3 (with respect to H_λ) we get $G_\lambda f\leq s$ for each $\lambda>0$ (the relation $H^U\geq H^U_\lambda$ shows that s is superharmonic relative to H_o). There results $G_o f\leq s$. On the other hand $G_o f-H^U G_o f=G^U f=s-H^U s$, i.e. $s-G_o f$ is balanced on E, which implies $s=G_o f$ because s is a potential. Now the statement of the theorem is proved for $\lambda=0$ and further it is not hard to deduce the full statement.

2.5. Corollary

If there exists a bounded continuous strict potential on X, p, (relative to H_o) then $P_t(C_o(X))\subset C_b(X)$ for each $t>0$. Moreover if $p\in C_o(X)$ then $P_t(C_o(X))\subset C_o(X)$ for each $t>0$.

IV. THE ADDITION OF LOCAL OPERATORS ON PRODUCT SPACES

1. A Simple Lemma

Let X be a locally compact space with a countable base and L a local operator on X. Suppose that L is locally dissipative and locally closed.

1.1. Lemma. Let U be a relatively compact open set such that $\partial U = \emptyset$ and

1° for any $x \in \partial U$ there exists a finite family $\{\varphi_1,..,\varphi_k\} \subset D(U,L)$ such that $\varphi_i > 0$, $L\varphi_i < -1$, $i=1,...,k$ and $\lim\limits_{\substack{y\to x\\ y\in U}} (\inf\limits_i \varphi_i(y)) = 0$,

2° there exists $\varphi \in D(U,L)$ such that $L\varphi < -1, \varphi \geqslant 0$ and $\|\varphi\| < \infty$,

3° the spaces $D_o(U)$ and $LD_o(U)$ are dense in $C_o(U)$, where

$$D_o(U) = \{f \in C_o(U) \cap D(U,L) / Lf \in C_o(U)\}.$$

Then U is P-regular.

Proof. First we are going to prove the following property:
(1) if $\{f_n / n \in N\}$ is a sequence in $D_o(U)$ such that $Lf_n \to 0$ uniformly on each compact subset of U and $\sup \|Lf_n\| < \infty$, then $f_n \to 0$ uniformly.

Let $\varepsilon > 0$. We choose a finite family $\{\varphi_1,..,\varphi_k\} \subset D(U,L)$ such that $\varphi_i > 0$, $L\varphi_i < -1, i=1,...,k$ and the set $K = \bigcap\limits_{i \leqslant k} \{\varphi_i \geqslant \varepsilon\}$ is compact. If $f \in D_o(U)$, $|Lf| \leqslant \varepsilon$ on K and $\|Lf\| \leqslant 1$, then from I.1.4 we deduce

$$\sup_{\overset{\circ}{K}} (f - \varepsilon\varphi) \leqslant \max (0, \sup_{\partial K} (f - \varepsilon\varphi)),$$

because $Lf - \varepsilon L\varphi > 0$ on K. Hence

$$\sup_{\overset{\circ}{K}} f \leqslant \varepsilon\|\varphi\| + \max (0, \sup_{\partial K} f).$$

On the other hand $L(f - \varphi_i) > 0$ implies $f \leqslant \varphi_i$, $i=1,...,k$, and hence $\sup\limits_{U \setminus \overset{\circ}{K}} f \leqslant \varepsilon$, which leads to

$$\|f\| \leqslant \varepsilon (\|\varphi\| + 1).$$

Property (1) results from this inequality.

Let now $f \in C_b(U)$. Condition 3° allows us to choose a sequence $\{\varphi_n / n \in N\} \subset D_o(U)$ such that $\sup\limits_n \|L\varphi_n\| < \infty$ and $L\varphi_n \longrightarrow f$ uniformly on the compact subsets of U. Next we assert that $\{\varphi_n\}$ is a Cauchy sequence in $C_o(U)$. If it is not, we have a $\delta > 0$ and a subsequence

$\{\varphi_{n_k}/k\in N\}$ such that $\|\varphi_{n_k}-\varphi_{n_{k+1}}\|\geq\delta$, $k\in N$. On the other hand $L\varphi_{n_k}-L\varphi_{n_{k+1}}\longrightarrow 0\,(k\to\infty)$ uniformly on the compact subsets of U and this contradicts (1). We conclude that $\varphi_n\to u$ uniformly, $u\in C_0(U)\cap D(U,L)$, and $Lu=f$. This proves the lemma.

1.2. Corollary. Let us suppose that the family of all P- and D-regular sets forms a base of X. If U,V are two P-regular sets then $U\cap V$ is P-regular too.

Proof. Conditions 1^0 and 2^0 from the lemma are obviously fulfilled. In order to check condition 3^0 one uses III.2.4.

2. The Sum of Two Local Operators

Let X_1, X_2 be locally compact spaces with countable bases. For $i=1,2$ let L^i be a locally dissipative local operator on X_i such that the family of all P- and D-regular sets is a base of X_i. We denote by $X=X_1\times X_2$; if $U\subset X$ and $x\in X_1$, $y\in X_2$ we put $U_x=\{z\in X_2/(x,z)\in U\}$, $U_y=\{z\in X_1/(y,z)\in U\}$. If f is a function on U, then $f_x=f(x,.)$ is defined on U_x and similarly $f_y=f(.,y)$ is defined on U_y. We define a local operator on X, $L=L(L^1,L^2)$, as follows: if U is an open set in X, then $D(U,L)$ is the family of all functions $f\in C(U)$ such that:

1^0 for any $x\in X_1$, $f_x\in D(U_x,L^1)$

2^0 for any $y\in X_2$, $f_y\in D(U_y,L^2)$

3^0 $Lf\in C(U)$, where $Lf(x,y)=L^1f_y(x)+L^2f_x(y)$, $(x,y)\in U$.

L is obviously locally dissipative. If U_i is an open set in X_i and $f_i\in D(U_i,L_i)$, $i=1,2$, then $f_1\otimes f_2$ $D(U_1\times U_2,L)$ and $L(f_1\otimes f_2)=L^1f_i\otimes f_2+f_1\otimes L^2f_2$. Thus one can prove the property from Corollary I.1.9. Then $\tilde{L}$ the local closure of L exists, is locally dissipative and locally closed. We denote by $L^1+L^2=\tilde{L}$. Next we are going to prove that the family of all P- and D-regular sets is a base of X.

2.1. Proposition. Let U_i be a P-regular set in X_i such that there exists $\varphi_i\in D(U_i,L^i)$, $\varphi_i\geq 1$, $L^i\varphi_i\leq 0$, $i=1,2$. Then $U=U_1\times U_2$ is P-regular (with respect to L^1+L^2).

Proof. We are going to apply Lemma 1.1. Thus we remark that

$\varphi_1 \otimes G^{U_2} 1$ fulfils the requirements of 1.1, 2°. Condition 1.1, 1° may be checked using the functions $\varphi_1 \otimes G^{U_2} 1$ and $G^{U_1} 1 \otimes \varphi_2$. In the reminder proof we are going to check 1.1, 3°. First we introduce some notations. We denote by $G^i_\lambda = G^{U_i}_\lambda$. The Hille-Yosida theorem applied for the space $C_o(U_i)$ gives us a C_o-class semigroup $\{P^i_t / t>0\}$ such that $G^i_\lambda = \int_0^\infty \exp(-\lambda t) P^i_t dt, \lambda \geq 0$. $\{P^i_t / t>0\}$ extend also as sub-Markov semigroup of kernels on U_i. The product semigroup $P_t = P^1_t \otimes P^2_t$ is the natural tensor product of kernels, i.e. if $(x,y) \in U$ then $P^{(x,y)}_t = P^{1,x}_t \otimes P^{2,y}_t$ is a product measure on U. $\{P_t / t>0\}$ is also a C_o-class semigroup on the space $C_o(U) = \overline{C_o(U_1) \otimes C_o(U_2)}$. Now we remark that $G_\lambda = \int_0^\infty \exp(-\lambda t) P_t dt, \lambda \geq 0$ define a family of kernels on U and $G_o 1 \leq G^1_o 1 \otimes 1$. If $f_i \in C_o(U_i)$, $i=1,2$, then

$$\left| \int_s^\infty P^1_t f_1 \otimes P^2_t f_2 dt \right| \leq \int_s^\infty P^1_t 1 dt \, \| f_1 \| \, \| f_2 \|$$

But $\int_s^\infty P^1_t 1 dt \to 0$ $(s \to \infty)$ uniformly because $\int_0^\infty P^1_t dt = G^1_o 1 \in C_o(U_1)$. Since $\int_0^s P^1_t f_1 \otimes P^2_t f_2 dt \in C_o(U)$ we deduce $G_o(f_1 \otimes f_2) \in C_o(U)$. Further the relation $\| G_o \| \leq \| G^1_o 1 \|$ shows that G_o is a linear operator on $C_o(U)$.

Now we remark that $G^1_o(C_o(U_1)) \otimes G^2_o(C_o(U_2)) \subset D_o(U)$. Namely if $f_i \in C_o(U_i)$, $i=1,2$, then $G^1_o f_1 \otimes G^2_o f_2 \in D(U,L)$ and $L(G^1_o f_1 \otimes G^2_o f_2) = G^1_o f_1 \otimes f_2 + f_1 \otimes G^2_o f_2$. This shows that $\overline{D_o(U)} = C_o(U)$. Further we need the follows equalities:

$$G_o(G^1_o f_1 \otimes G^2_o f_2)(x_1,x_2) = G^1_o(G_o(f_1 \otimes G^2_o f_2)(\cdot,x_2))(x_1) =$$

(1) $$= G^2_o(G_o(G^1_o f_1 \otimes f_2)(x_1,\cdot))(x_2), \quad f_i \in C_o(U_i), x_i \in U_i, \ i=1,2,$$

(2) $$G^1_o f_1 \otimes G^2_o f_2 = G_o(G^1_o f_1 \otimes f_2 + f_1 \otimes G^2_o f_2), \quad f_i \in C_o(U_i), \ i=1,2.$$

The first results by strightforward computations. The second equality results from

$$\int_0^\infty \int_0^\infty (P^1_s f_1)(P^2_t f_2) ds dt = \int_0^\infty \left(\int_t^\infty (P^1_u f_1)(P^2_t f_2) du \right) dt + \int_0^\infty \left(\int_s^\infty (P^1_s f_1)(P^2_u f_2) du \right) ds.$$

From (1) and (2) we deduce $L(G_o(G^1_o f_1 \times G^2_o f_2)) = G^1_o f_1 \times G^2_o f_2$, and hence $G^1_o(C_o(U_1)) \otimes G^2_o(C_o(U_2)) \subset LD_o(U)$, $\overline{LD_o(U)} = C_o(U)$. The operator $L^1 + L^2$ being an extension of L we conclude that 1.1, 3° is fulfilled an U is P-regular.

If the set U_i satisfies $\overline{U}_i \subset V_i$ for some P-regular set, V_i, then $\varphi_i = \alpha G^{V_i} 1$ fulfils $\varphi_i \geq 1$ on U_i for suitable α. Then we deduce that the sets $U = U_1 \times U_2$ which fulfil the requirements from Proposition 2.1 form a base of X. Hence $L_1 + L_2$ has a base of P-regular sets or (equivalently on account of I.1.12.5°) it has a base of P-and D-regular sets.

Now let X_1, X_2, X_3 be three locally compact spaces with countable bases. For $i=1,2,3$, let L^i be a locally dissipative local operator on X_i such that family of P-and D-regular sets is a base of X_i. On $X_1 x X_2 x X_3$ we define a local operator, L^o, as follows: if U is an open set in $X_1 \times X_2 \times X_3$ then $D(U, L^o)$ is the family of all functions $f \in C(U)$ such that for each $x = (x_1, x_2, x_3) \in U$,

a) $f_{(x_i, x_j)} \in D(U_{(x_i, x_j)}, L_k)$ for $i \neq j \neq k \neq i$, $i,j,k \in \{1,2,3\}$,

b) $L^o f \in C(U)$, where $L^o f(x_1, x_2, x_3) = L_1 f_{(x_2, x_3)}(x_1) + L_2 f_{(x_1, x_3)}(x_2)$
$$+ L_3 f_{(x_1, x_2)}(x_3).$$

L^o is locally dissipative and has the property from Corollary I.1.9. Hence its local closure $\widetilde{L}^o$ exists. We also note that the proof of 2.1 may be repeated here word by word, and hence we deduce that $\widetilde{L}^o$ has a base of P- and D-regular sets. On the other hand from $L^1 + L^2$ and L^3 we get another local operator $L^{oo} = L(L^1 + L^2, L^3)$ on $(X_1 \times X_2) \times X_3$ defined by 1°, 2°, 3°, such that its local closure $\widetilde{L}^{oo}$, is $(L^1 + L^2) + L^3$. It is easy to see that $D(U, L^o) \subset D(U, L^{oo})$ for each open set U and L^{oo} extends L^o and so $\widetilde{L}^{oo}$ extends $\widetilde{L}^o$. If U is P- regular or D-regular with respect to $\widetilde{L}^o$ then it is alike with respect to $\widetilde{L}^{oo}$ and the kernels H^U, G^U associated to $\widetilde{L}^o$ coincide with those associated to $\widetilde{L}^{oo}$. If U is P- and D-regular and $\varphi \in D(V, \widetilde{L}^{oo})$, $\overline{U} \subset V$ $\|L^{oo}\varphi\| < \infty$, then $G^U(-\widetilde{L}^{oo}\varphi) + H^U \varphi = \varphi$. Hence $\varphi \in D(U, \widetilde{L}_o)$ and $\widetilde{L}^o \varphi = \widetilde{L}^{oo} \varphi$. Further one deduces $\widetilde{L}^o = \widetilde{L}^{oo}$. Thus we may put the notation $L^1 + L^2 + L^3 = (L^1 + L^2) + L^3 = L^1 + (L^2 + L^3)$ and conclude that the addition of local operators is well defined for each finite family.

<u>Remark.</u> One can define the same operator $L^1 + L^2$ by replacing condition 3° in the definition of $L(L^1, L^2)$ with the stronger one:

$$(x,y) \to L^1 f_y(x) \in C(U),$$
$$(x,y) \to L^2 f_x(y) \in C(U).$$

<u>Note</u>. Relation (1) from above was comunicated to us by N.Boboc.

3. The Sum of a Series of Local Operators

Let $\{X_i/i\in N\}$ be a sequence of compact spaces with countable bases and for each $i\in N$ let L^i be a locally dissipative local operator on X_i such that $1\in D(X_i,L^i)$, $L^i 1=0$ and the family of all P-and D-regular sets of X_i is a base. We know from Corollary III.2.3 that there exists a Markov semigroup of kernels $\{P_t^i/t>0\}$ on X_i which is also a C_o-class semigroup of contractions on the Banach space $C(X_i)$ whose infinitesimal generator has $D(X_i,L^i)$ as domain and concides with L^i as linear operator on this last space. We denote by $X=\prod_{i\in N} X_i$. For $J\subset N$ we put $X(J)=\prod_{i\in J} X_i$. If $A\subset X$, $f:A\longrightarrow R$ and $x\in X(J)$ we put $A_x=\{y\in X(N\setminus I)/(x,y)\in A\}$ and $f_x:A_x\longrightarrow R$, $f_x(.)=f(x,.)$. If J is finite we put $L(J)=\sum_{i\in J} L^i$.

We define a local operator, L, on X as follows: if U is an open set $U\subset X$, then $D(U,L)$ is the family of all functions, f, which fulfil the following property:

For each point $x_o\in U$ there exist a finite set $J=J(f,x_o)\in N$, two open sets U_1,U_2, such that $U_1\subset X(J)$, $U_2\subset X(N\setminus J)$, $x_o\in U_1\times U_2\subset U$ and a function $g\in D(U_1,L(J))$ such that $f(x_1,x_2)=g(x_1)$ for each $(x_1,x_2)\in U_1\times U_2$.

For a function f and (x_1,x_2) as above we define $Lf(x_1,x_2)=L(J)g(x_1)$. L is locally dissipative and fulfis the property from Corollary I.1.9. Its local closure $\tilde{L}$ is locally dissipative and locally closed. We denote by $\sum_{i\in N} L^i=\tilde{L}$. The next proposition implies that $\sum_{i\in N} L^i$ has a base of P- and D-regular sets.

3.1. Proposition. Let J be a finite subset of N and U a P-regular set in $X(J)$ (with respect to $L(J)$). Then $U_o=U\times X(N\setminus J)$ is P-regular (with respect to $\sum_{i\in N} L^i$).

Proof. The proof is similar to that of Proposition 2.1, therefore we only sketsch it. Let $\{P_t^*/t>0\}$ be the semigroup of kernels on U such that $G_\lambda^U=\int_0^\infty \exp(-\lambda t)P_t^* dt$, $\lambda\geq 0$. For a fixed finite set $K\subset N\setminus J$ we define $P_t^K=P_t^*\otimes(\bigotimes_{i\in K} P_t^i)$, $G_\lambda^K=\int_0^\infty \exp(-\lambda t)P_t^K dt$, $\lambda\geq 0$ and for $i\in K$ put $G_\lambda^i=\int_0^\infty \exp(-\lambda t)P_t^i dt$, $\lambda>0$. First we are going to prove that $U_K=U\times X(K)$ is P-regular with respect to $L(J\cup K)$. Since $G_o^i=\sup_{\lambda>0} G_\lambda^i$ is not finite for $i\in K$ we have no equality analogous to (1) or (2) from 2.1. Therefore we consider $\lambda>0$ and put $\alpha=\lambda/n$ where n is the number of elements from K. Then for $f^*\in C_o(U)$ and $f_i\in C(X_i)$, $i\in K$ we have

$g=G_o^U t^* \otimes (\bigotimes_{i\in K} G_\alpha^i f_i)\in D(U_K, L(J\cup K))$ and $L(J\cup K)_\lambda\ g=(\sum_{i\in J} L^i+\sum_{i\in K} L_\alpha^i)g\in C_0(U_K)$.

For $L(J\cup K)_\lambda$ we have some equalities similar to (1) and (2) from 2.1, namely

$$(1)\quad G_\lambda^K(G_o^U\otimes(\bigotimes_{i\in K} G_\alpha^i))=(G_o^U\otimes I_{C(X(K))})G_\lambda^K(I_{C_o(U)}\otimes(\bigotimes_{i\in K} G_\alpha^i)) =$$

$$=(I_{C_o(U)}\otimes G_\alpha^j\otimes I_{C(X(K\setminus\{j\}))})G_\lambda^K(G_o^U\otimes I_{C(X_j)}\otimes(\bigotimes_{\substack{i\in K\\ i\neq j}} G_\alpha^i)),$$

$$(2)\quad G_o^U\otimes(\bigotimes_{i\in K} G_\alpha^i) = G_\lambda^K(I_{C_o(U)}\otimes(\bigotimes_{i\in K} G_\alpha^i) + \sum_{j\in K} G_o^U\otimes I_{C(X_j)}\otimes(\bigotimes_{\substack{i\in K\\ i\neq j}} G_\alpha^i)$$

One deduces $L(J\cup K)_\lambda G_\lambda^K g=-g$ and from Lemma 1.1 there results that U_k is P-regular with respect to $L(J\cup K)_\lambda$, just like in the proof of 2.1. Further one deduces that U_K is P-regular with respect to $L(J\cup K)$ by using the result from I.1.12.3°.

Further one deduces that U_o is P-regular relative to $\sum_{i\in N} L^i$ by applying Lemma 1.1 again.

4. The Operator L-d/dt

Let X be a locally compact space with a countable base and L a locally dissipative local operator such that the family of all P- and D-regular sets with respect to L is a base. In this section we shall construct another local operator, L-d/dt, on the space X × R, which fulfils the same properties. It should be noted that J.P.Roth [34] ch.IV constructed a similar operator within a different framework.

We denote by $T=(0,1]$ the torus and consider it as a differentiable manifold. On the space T x R we consider L^o, the local closure of $\partial^2/\partial x^2-d/dt$ ($x\in T$, $t\in R$). L^o is locally dissipative, has a base of P- and D-regular sets (see I.3.3) and is translation invariant. The sum $L+L^o$ is also locally dissipative and has a base of P- and D-regular sets. We shall use the following property: if U is an open set in X x T x R, then

(i) $f\in D(\tau_x U,\ L+L^o)$ iff $f\circ\tau_{-x}\ D(U,L+L^o)$ and

$L+L^o f=(L+L^o(f\circ\tau_x))\circ\tau_{-x}$, for each $x\in T$,

where $\tau_x: X\times T\times R\to X\times T\times R$ is defined by $\tau_x(z,y,s)=(z,x+y,s)$. We denote by $\theta: X \times T \times R \to X \times R$, the map defined by $\theta(z,x,s)=(z,s)$. Then a local operator denoted by $\check{L}$ is defined $X \times R$: if $U \subset X \times R$ is an open set, a function f belongs to $D(U,\check{L})$ if and only if $f\circ\theta \in D(\theta^{-1}(U)),$ $L+L^{o})$ and $\check{L}f=g$, where g is the unique function in $C(U)$ such that $L+L^{o}(f\circ\theta)=g\circ\theta$ (the existence of g is a consequence of relation (1)).

4.1. <u>Proposition</u>. Let $W\subset X$ be a P-regular set (with respect to L) and $h\in D(W,L)$ such that $Lh=0, \alpha\leq h\leq 1/2$ on W for some $\alpha>0$. For $t_o\in R$ we define $p: W \times R \to R$, by $p(x,t)=(t-t_o)G^{W}1(z)-(t-t_o)^2 h(z)$, and put $U=\{(x,t)\in W\times R / p(x,t)>0\}$. Then U is P-regular with respect to $\check{L}$.

<u>Proof.</u> We have $p\circ\theta\in D(\theta^{-1}(U), L+L^{o})\cap C_o(\theta^{-1}(U))$ and $L+L^{o}(p\circ\theta)$ $(z,x,t)=-G^{W}1(z)-(t-t_o)(1-2h(z))<0$ on $\theta^{-1}(U)$. Moreover for each neighborhood A of the compact set $\partial W\times T\times\{t_o\}$ there exists $a\in R$, $a>0$ such that $L+L^{o}(p\circ\theta)<-a$ on $\theta^{-1}(U)\setminus A$. Now we may use Theorem III.2.4.2° and get a kernel V on $\theta^{-1}(U)$ such that

$$Vf\in C_o(\theta^{-1}(U)\cap D(\theta^{-1}(U), L+L^{o}), \quad L+L_o Vf=-f, \text{ for each}$$

$f\in C_b(\theta^{-1}(U))$ provided $\overline{\text{supp } f}\cap(\partial W \times T \times \{t_o\})=\phi$

If $f\in C_b(\theta^{-1}(U))$, $0\leq f\leq 1$, and $\overline{\text{supp } f}\cap(\partial W\times T\times\{t_o\})=\phi$, then $L+L_o G^{W}1=-1\leq -f=L+L_o Vf$ and $Vf=0$ on $\partial\theta^{-1}(U)$. It follows $Vf(z,x,s)\leq G^{W}1(z)$ for each $(z,x,s)\in\theta^{-1}(U)$.

Further let $\{\varphi_n / n\in N\}\subset C_c(W)$ be a sequence such that $0\leq\varphi_n\leq\varphi_{n+1}\leq 1$ and for any compact set $K\subset W$ there exists $n\in N$ such that $\varphi_n=1$ on K. If $f\in C(\theta^{-1}(U))$, $0\leq f\leq 1$, and $m\geq n$, then $0\leq V((\varphi_m-\varphi_n)f)\leq$ $\leq\sup\{V((\varphi_m-\varphi_n)f)(z,x,s) / (z,x,s)\in\theta^{-1}(U),\ \varphi_n(z)<1\}\leq$

$$\leq\sup\{G^{W}1(z)/\varphi_n(z)<1\}\to 0 \qquad (n\to\infty).$$

We deduce that $V(\varphi_n f)\to Vf$ uniformly and $Vf\in C_o(\theta^{-1}(U))$. Hence $V_o f\in D(\theta^{-1}(U), L+L^{o})$ and $L+L^{o}Vf=-f$. Thus $\theta^{-1}(U)$ is P-regular with respect to $L+L^{o}$ and $V=G^{\theta^{-1}(U)}$. Further on account of (1), straightforward computations show that U is P-regular with respect to $\check{L}$ and G^{U} satisfies

$$(2) \qquad (G^{U}f)\circ\theta=V(f\circ\theta) \quad \text{for each} \quad f\in C_b(U).$$

4.2. Corollary

$\check{L}$ has a base of P- and D-regular sets.

Proof. Let $x \in X$ and $\{V_n/n\in N\}$ a sequence of D-regular neighbourhoods which tends to $\{x\}$. Then relation (3) from I.4.3 shows that $H^{V_n}1(x) \to 1$. Thus for large n, $H^{V_n}1=h$ satisfies $h>0$ on a neighbourhood of x. Then by Proposition 4.1 we can construct a base of sets similar to U.

Next we define $\hat{L}$, another local operator on X x R: if U is an open set in X x R, then D(U,L) is the family of all functions $f \in C(U)$ which fulfil

1° for each $x\in X$, $f_x \in C^1(U_x)$,

2° for each $s\in R$, $f_s \in D(U_s,L)$,

3° $\hat{L}f \in C(U)$, where $\hat{L}f(x,s)=Lf_s(x)-d/dt f_x(s)$, $(x,s)\in X \times R$.

Once again the property from Corollary I.1.9 is easy to check. We denote by L-d/dt the local closure of $\hat{L}$. Obviously $\check{L}$ extends $\hat{L}$. Since $\check{L}$ is locally closed there results that it extends L-d/dt.

4.3. Proposition

$\check{L}$=L-d/dt

Proof. Let $U\subset X$ be a P-regular set with respect to L and let $\{P_t/t>0\}$ the semigroup of kernels on U such that $G^U_\lambda=\int_0^\infty \exp(-\lambda t)P_t dt$, $\lambda\geq 0$. We define on U x R a semigroup $\{Q_t/t>0\}$ by putting $Q_t f(x,s)=$ $=(P_t(f(.,s-t))(x)$, for each $f \in C_b(U \times R)$, i.e. $Q_t=P_t \otimes P'_t$, where $\{P'_t/t>0\}$ is the left-translation semigroup. Further $G=\int_0^\infty Q_t dt$ is a kernel on U x R and $G1(x,s)=G^U 1(x)$.

Let $f\in C_o(U)$ and $\varphi\in C^1(R)\cap C_c(R)$. Then we have

$G(G^U f \otimes \varphi)(x,s)=G^U(G(f \otimes \varphi)(.,s))(x)$ for $(x,s) \in U \times R$ and

$G(G^U f \otimes \varphi)(x,.) \in C^1(R)$, $d/dt\, G(G^U f \otimes \varphi)(x,.)=G(G^U f \otimes \varphi')(x,.)$.

An equality analogous to relation (2) from 2.1 holds and it allows us to deduce that, $G(G^U f \otimes \varphi)$ $D(U \times R,\hat{L})$ and $\hat{L}G(G^U f \otimes \varphi)=-G^U f \otimes \varphi$.

On the other hand, since $P_t G^U 1 \to 0$ uniformly when $t \to \infty$, one deduces that $G(G^U f \otimes \varphi) \in C_o(U \times R)$. Thus $Gf \in D(U \times R, L-d/dt) \cap C_o(U \times R)$ and $(L-d/dt)Gf=-f$ for each $f \in C_o(U \times R)$.

Let $\{\varphi_n\}$ be a sequence in $C_o(U \times R)$ such that $0 \leq \varphi_n \leq \varphi_{n+1} \leq 1$ and $\lim_{n\to\infty} \varphi_n = 1$ on $U \times R$. The relation $G1 = \lim_{n\to\infty} G\varphi_n$ implies that $G(1-\varphi_n) \to 0$ uniformly on each compact subset of $U \times R$. If $f \in C_b(U \times R)$ then $G(f\varphi_n) \to Gf$ uniformly on each compact subset of $U \times R$, which shows that $Gf \in D(U \times R, L-d/dt)$ and $(L-d/dt)Gf=-f$.

Next, for each open set $V \subset X \times R$ we denote by $\mathcal{H}(V) = \{f \in D(V, \check{L}) / \check{L}f = 0\}$. Then from Corollary 4.2 we deduce that $\mathcal{H} = \{\mathcal{H}(V)/V$ open set$\}$ defines a quasiharmonic space on $X \times R$. From Proposition II.1.4 we deduce that Gf is a potential for each $f \in C_{o+}(U \times R)$. From Proposition II.1.2, 2° it follows that $G1 = \sum_{n\in N} G(\varphi_{n+1} - \varphi_n)$ is also a potential Then Theorem II.2.8 shows that $Gf = f.G1$ for each $f \in C_b(U \times R)$. Finally we deduce that $\check{L} = L - d/dt$ by using Theorem II.6.4 b).

5. Bauer Spaces and Strong Feller Semigroups

Let X be a locally compact space with a countable base and L a local operator which is locally dissipative and suppose that the family of all P- and D-regular sets is a base of X. We denote by $\mathcal{H} = \{\mathcal{H}(U)/U$ open set$\}$, where $\mathcal{H}(U) = \{f \in D(U,L)/Lf=0\}$ for each open set $U \subset X$. Then $(X, \mathcal{H})$ is a quasiharmonic space.

Now we recall that a kernel R on a locally compact space, T, is said to be <u>strong Feller</u> if $Rf \in C(T)$ for each $f \in \mathcal{B}_b(T)$.

We say that L satisfies the property of "strong Feller resolvent" if:

(SFR) There exists a covering $\mathcal{U}$ of P-regular sets such that G^U is strong Feller for each $U \in \mathcal{U}$.

We say that L satisfies the property of "strong Feller semigroup " if:

(SFS) There exists a covering $\mathcal{U}$ of P-regular sets such that if $U \in \mathcal{U}$ and $\{P_t^U / t>0\}$ is the semigroup of kernels on U which fulfils $G_\lambda^U = \int_0^\infty \exp(-\lambda t) P_t^U dt$, for each $\lambda > 0$, then each kernel P_t^U, $t>0$ is strong Feller.

Further we recall some results related to the strong Feller property.

5.1. <u>Lemma</u>. Let T be a locally compact space with countable base and V_1 , V_2 two sub-Markov kernels on T which are strong Feller. Then the family of functions $\{V_1V_2f/f\in B(T),\ |f|\leq 1\}$ is equicontinuous. For a proof of this Lemma see [28].

5.2. <u>Proposition</u>. The following properties are equivalent:

1° L satisfies (SFR),

2° $(X,\mathcal{H})$ is a Bauer space in the meaning of [13],

3° For each P-regular set, U, the kernels G_λ^U, $\lambda\geq 0$ are strong Feller.

4° The global resolvent $\{G_\lambda/\lambda>0\}$ (i.e. the resolvent of the global process constructed in Theorem I.2.5) is strong Feller.

<u>Proof</u>. $4^o\Longrightarrow 3^o$ It results from the relation $G_\lambda^U=G_\lambda-H_\lambda^U G_\lambda$.

$3^o\Longrightarrow 2^o$ Since 1 is excessive, for each $x\in X$ we deduce from Corollary I.4.3 that $H^U 1(x)>0$, for suitable open neighbourhood of x. This shows that $\mathcal{H}$ is non-degenerate at any point of X. Now let $\{h_n\}\subset\mathcal{H}(U)$ be a sequence such that $0\leq h_n\leq h_{n+1}\leq 1$. In order to show that $h=:\lim h_n\in\mathcal{H}(U)$ we may assume there exists $h'\in\mathcal{H}(U)$ such that $h\leq h'$ and suppose U to be P-regular. Since h and h'-h are excessive with respect to the resolvent $\{G_\lambda^U\}$, they are lower semicontinuous. Their sum h+h'-h=h' is continuous, and hence h must be continuous.

The reminder proof is obvious.

The main result of this section is the following theorem.

5.3. <u>Theorem</u>. The following properties are equivalent:

1° L satisfies (SFS) .

2° L-d/dt satisfies (SFR) .

3° If X^o is a locally compact space with a countable base and L^o is a locally dissipative local operator such that the family of all P- and D-regular sets is a base of X^o and L^o satisfies (SFR), then $L+L^o$ satisfies (SFR) too.

4° If $X^o=T\times R$ and L^o is the local closure of $\partial^2/\partial x^2-\partial/\partial t$ (as in section 4), then $L+L^o$ satisfies (SFR).

5° If U is an open set, $U\subset X$, and $\{P_t/t>0\}$ is a sub-Markov semigroup of kernels which is also a (C_o)-class semigroup of operators on a Banach space $F\subset C_b(U)$ whose infinitesimal generator Δ, has a domain,

$D(\Delta)$, such that $D(\Delta) \subset D(X,L)$, $\Delta = L$ as linear operators on $D(\Delta)$, and $C_0(U) \subset \overline{D(\Delta)} = F$, then P_t is strong Feller for each $t>0$.

In order to prove this theorem we need the following lemma, which was independently proved by three authors: E.Popa [33] (but not explicitely stated), U.Schirmeier [36], §2.Satz and L.Stoica [38] Proposition VI.1.4.

5.4. Lemma. Let X_1, X_2 be two locally compact spaces with countable bases and $\{P_t^i / t>0\}$ a sub-Markov semigroup of kernels on X_i such that $t \to P_t^i f(x)$ is right continuous for each $x \in X_i$, $f \in C_b(X_i)$. Suppose that $G_0^i = \int_0^\infty P_t^i dt$, $i=1,2$ are (finite) kernels and P_t^1, $t>0$, G_0^2 are strong Feller.

Then $G = \int_0^\infty P_t^1 \otimes P_t^2 dt$ is a strong Feller kernel on $X_1 \times X_2$.

Proof. From Lemma 5.1 there results that $\{P_t^1 f / f \in B(X_1), |f| \leq 1\}$ is an equicontinuous family for each $t>0$.

First we are going to prove that for a given $f \in B(X_1 \times X_2)$ such that $|f| \leq 1$ and a given compact set $K \subset X_2$ the family

$$\{Gf(\cdot, y_2) / y_2 \in K\} \quad \text{is equicontinuous.}$$

If $y_2 \in X_2$, then μ^{y_2} will denote the measure on $X_2 \times R_+$ which fulfils

$$\int g(x_2,t)\, d\mu^{y_2}(x_2,t) = \int_0^\infty P_t^2(g(\cdot,t))(y_2)\, dt \ , \qquad g \in B_b(X_2 \times R_+) \ .$$

Since $\mu^{y_2}(1) = G_0^2 1(y_2)$ we deduce that there exists a real number $c>0$ such that $\mu^{y_2}(1) \leq c$ for each $y_2 \in K$. Let $y_1 \in X_1$ and $\varepsilon>0$. The family $\{P_t^1 f_{x_2} / x_2 \in X_2 \ , \ t \geq \varepsilon/4\}$ being equicontinuous we can choose a neighbourhood W of y_1 such that

$$|P_t^1 f_{x_2}(y_1) - P_t^1 f_{x_2}(y_1')| < \varepsilon/2c$$

for each $y_1' \in W$, $x_2 \in X_2$ and $t \geq \varepsilon/4$. This implies

$$|G_0 f(y_1,y_2) - G_0 f(y_1',y_2)| \leq \varepsilon/4 + \varepsilon/4 + \int_{\{t \geq \varepsilon/4\}} |P_t^1 f_{x_2}(y_1) - P_t^1 f_{x_2}(y_1')|\, d\mu^{y_2}(x_2,t) \leq \varepsilon$$

for each $y_1' \in W$ and $y_2 \in K$, which proves our assertion.

Next we are going to prove that for $f \in B(X_1 \times X_2)$, $|f| \leq 1$, and

$y_1 \in X_1$ the function $G_0 f(y_1,.)$ is continuous on x_2. The properties of the semigroup $\{P_t^1/t>0\}$ and the result 10.VIII from [27] allow us to construct a finite measure, μ, on X_1 and a function $g \in \mathcal{B}((0,\infty) x X_1)$ such that

$$P_t^{1,y_1} = g(t,.).\mu \quad \text{for each } t>0.$$

If we put $h(x_1, y_2) = \int_0^\infty g(t,x_1) P_t^2 f_{x_1}(y_2)dt$, $x_1 \in X_1$, $y_2 \in X_2$, then

(1) $$Gf(y_1,y_2) = \int_0^\infty h(x_1,y_2)d\mu(x_1) \ .$$

Now we assert that $h(x_1,.)$ is continuous provided $\int_0^\infty g(t,x_1)dt<\infty$. Let $\varepsilon>0$. We choose $\varphi \in C_0((0,\infty))$ such that

$$\int_0^\infty |g(t,x_1)-\varphi(t)|dt<\varepsilon$$

and a finite number of constants α_1 , $\alpha_2,\ldots,\alpha_n>0$, and $c_1,c_2,\ldots,c_n$ such that

$$|\varphi(t) - \sum_{i=1}^n c_i \exp(-\alpha_i t)|<\varepsilon \qquad \text{on } R_+ .$$

Then

$$|h(x_1,y_2) - \sum_{i=1}^n c_i G_{\alpha_i}^2 f_{x_1}(y_2)| \leq \varepsilon + \varepsilon \int_0^\infty P_t^2 1(y_2)dt = \varepsilon + \varepsilon G^2 1(y_2) \ ,$$

where $G_\alpha^2 = \int_0^\infty \exp(-\alpha t)P_t^2 dt$, $\alpha \geq 0$ are the kernels of the resolvent associated to $\{P_t^2/t>0\}$. Since G_0^2 is strong Feller one deduces that for each $\alpha \geq 0$ the kernel G_α^2 is strong Feller. For a compact set $K \subset X_2$, there exists a constant c such that $G^2 1(y_2) \leq c$, $y_2 \in K$. We deduce that $h(x_1,.)$ can be uniformly approximated on K by continuous functions of the form $\sum_{i=1}^n c_i G_{\alpha_i}^2 f_{x_1}$. Hence $h(x_1,.)$ is continuous on X_2 .

Since $\int_0^\infty P_t^{1,y_1}dt$ is a finite measure on X_1 we deduce $\int_0^\infty g(t,.)dt \in L^1(\mu)$, and hence $\int_0^\infty g(t,.)dt<\infty$, μ-a.e. Since $|h(.,y_2)| \leq$ $< \int_0^\infty g(t,.)dt$ for each $y_2 \in X_2$ one deduces from relation (1) that $G_0 f(y_1,y_2^n) \to G_0 f(y_1,y_2^0)$, provided $y_2^n \to y_2^0$, i.e. $G_0 f(y_1,.)$ is continuous on X_2. Further it is easy to deduce that $G_0 f$ is continuous.

Proof of Theorem 5.4. $1^\circ \Rightarrow 3^\circ$ The proof of Proposition 2.1 and Lemma 5.4 show that the kernel G^U is strong Feller if $U=U_1 \times U_2$, $U_1 \subset X$ is P-regular and the kernels of its associated semigroup are strong Feller, and $U_2 \subset X^\circ$ is P-regular and G^{U_1} is strong Feller.

$3^\circ \Rightarrow 4^\circ$ It is obvious.

$4^\circ \Rightarrow 2^\circ$ The kernel G^U associated to the open set U from Proposition 4.1 is strong Feller. This results from relation 4.1 (2).

$2^\circ \Rightarrow 5^\circ$ Let $f \in D(\Delta)$ and define $\varphi : U \times R_+ \longrightarrow R$ by

(2) $$\varphi(x,t)=P_t f(x) \quad .$$

Then

(3) $$\varphi \in D(U\times(0,\infty), L-d/dt) \text{ and } (L-d/dt)\varphi=0 \quad .$$

Since $C_0(U) \subset \overline{D(\Delta)}$ and $L-d/dt$ is locally closed one deduces that (3) is still valid when φ is defined by (2) with $f \in C_0(U)$. Further a monotone class theorem together with Proposition 5.1, 2° shows that (3) is still valid when φ is defined by (2) with $f \in B_b(U)$. Particularly $P_t f \in C(U)$ for each $f \in B_b(U)$ and $t>0$.

$5^\circ \Rightarrow 1^\circ$ It is obvious.

Example

Let $X=(0,\infty)\times(-\infty,\infty)$ and let L^1 be the local closure associated to $(\partial/\partial x_1)^2 + x_1 \partial/\partial x_2$. Then L^1-d/dt satisfies (SFR) on account of Proposition 5.3 and of Corollary from [9] p.101, and hence L^1 satisfies (SFS). However ker L is not a Brelot space. If we denote by L^2 the local closure associated to $(1/x_1)(\partial/\partial x_1)^2 + \partial/\partial x_2$ we see that L^2 does not satisfy (SFS) while ker L^1=ker L^2.

It would be interesting to describe intrinsically the class of those harmonic spaces which have the property taht all associated local operators satisfy (SFS). For example the results of J.M. Bony [9] show that this property coincides (roughly speaking) with Brelot's axiom of convergence, providing the harmonic space is associated to a hypoelliptic differential operator in R^n. We do not know anything in the general case.

6. (SFS) for the Sum of a Series of Local Operators

Let $\{X_i/i\in N\}$ be a sequence of compact spaces with countable bases. Suppose that for each $i\in N$, L^i is a locally dissipative local operator on X_i such that $1\in D(X_i,L^i)$, $L^i\,1=0$, there exists a base of P- and D-regular sets, and L^i satisfies (SFS). In this section we are going to prove that $\sum_{i\in N} L_i$ satisfies (SFS) under suitable circumstances. We are going to use the notation from section 3. For a kernel V defined on a compact space, T, we put

(1) $$M(V)=\sup\{|Vf(x)-Vf(y)|/x,y\in T,\quad f\in \mathcal{B}_b(T),\ |f|\leq 1\}$$

Obviously we have $M(V)\leq 2$, provided B is sub-Markov.

6.1. Lemma. Suppose that P_n is a sub-Markov kernel on X_n such that $P_n 1>0$ and the family $\{P_n f/f\in\mathcal{B}(X_n),\ |f|\leq 1\}$ is equicontinuous, for each $n\in N$. If $\sum_{n\in N} M(P_n)<\infty$, then $P=\bigotimes_{n\in N} P_n$ is a kernel on X such that $\{Pf/f\in\mathcal{B}(X_n),\ |f|\leq 1\}$ is equicontinuous.

Proof. Let $x=(x_0,x_1,\dots)\in X$ and $\varepsilon>0$. We choose $n_0\in N$ such that $\sum_{n\geq n_0} M(P_n)<\varepsilon/2$ and for each $k<n_0$ we choose V_k a neighbourhood of x_k such that

$$|P_k f(x_k)-P_k f(y_k)|<\varepsilon/2n_0\ ,\quad \text{if } y_k\in V_k\ ,\ f\in\mathcal{B}(X_k),\quad |f|\leq 1.$$

We are going to prove that, for $y\in \prod_{k=0}^{n_0-1} V_k\times \prod_{i=n_0}^{\infty} X_i$, $f\in\mathcal{B}(X)$, $|f|\leq 1$, we have

(2) $$|Pf(x)-Pf(y)|<\varepsilon\ .$$

Let $y=(y_0,y_1,\dots)\in X$, with $y_k\in V_k$ for $k<n_0$. We put $z_k=(x_0,x_1,\dots,x_k,\ y_{k+1},y_{k+2},\dots)$, for each $k\in N$. If $f\in C(\prod_{i=0}^{p} X_i)$, $p\in N$ is regarded as a function $f\in C(X)$, then $Pf=(\bigotimes_{i=1}^{p} P_i)f\otimes(\bigotimes_{i=p+1}^{\infty} P_i 1)$.
Then for $k>p$ we have

$$Pf(z_k)=(\bigotimes_{i=1}^{p} P_i)f(x_1\dots x_p)\cdot\prod_{i=p+1}^{k} P_i 1(x_i)\ \prod_{i=k+1}^{\infty} P_i 1(y_i)\ .$$

On account of Lemma 6.2 (stated below) we deduce

$$\lim_{k\to\infty} Pf(z_k)=Pf(x) \ .$$

On the other hand, if $|f|\leq 1$, one deduces

$$|Pf(z_k)-Pf(z_{k+1})|<\varepsilon/2n_0 \ , \quad \text{for } k<n_0-1 \ ,$$

$$|Pf(z_k)-Pf(z_{k+1})|<M_k \ , \text{ for } k\geq n_0-1 \ .$$

Thus we get $|Pf(y)-Pf(x)|<\varepsilon$. Further it is easy to deduce relation (2) for each $f\in C(X)$ such that $|f|\leq 1$ and using a monotone class theorem to deduce the relation for $f\in\mathcal{B}(X)$ such that $|f|\leq 1$.

6.2. Lemma. Let $0<a_n$, $b_n\leq 1$, $n\in N$, be such that $\sum_{n\in N}|a_n-b_n|<\infty$. Then $\lim_{k\to\infty}\prod_{n=k}^{\infty} a_n=\lim_{k\to\infty}\prod_{n=k}^{\infty} b_n$.

6.3. Theorem. If the semigroups $\{P_t^i/t>0\}$, $i\in N$ satisfy the condition $\sum_{i\in N} M(P_t^i)<\infty$ for each $t>0$, then $\sum_{i\in I} L^i$ satisfies (SFS).

Proof. Since the infinitesimal generator of $\{P_t^i/t>0\}$ is L^i considered as a linear operator on $D(X_i,L^i)$ by Theorem 5.4, 5° one deduces that P_t^i is strong Feller.

Further Lemma 5.1 and Lemma 6.1 allow us to deduce that the kernels $P_t^{U_o}$, $t>0$ defined by $P_t^{U_o}=P_t^*\otimes(\bigotimes_{i\in N\setminus J} P_t^i)$, where U_o and P_t^* have the meaning from 3.1, are strong Feller. This implies (SFS) for $\sum_{i\in N} L_i$

Next we are going to show that Theorem 6.3 applies for a large class of examples.

Lemma 6.4. Let T be a compact space with a countable base and let $\{P_t/t>0\}$ be a Markov semigroup of kernels.

1° For any $s,t>0$,

$$M(P_{s+t})\leq(1/2)M(P_s)M(P_t)\leq M(P_s) \ .$$

2° If the kernels P_t, $t>0$ are strong Feller and there exists a measure μ on T such that $P_t^x\to\mu\,(t\to\infty)$ for each $x\in T$, then $\lim_{t\to\infty} M(P_t)=0$.

Proof

1° For $f \in \mathcal{B}_b(T)$, $|f| \leq 1$ we put $\alpha = (1/2)(\sup P_s f + \inf P_s f)$ and $g = (2/M(P_s))(P_s f - \alpha)$. Since $P_t \alpha = \alpha$ we deduce

$$\sup P_{s+t} f - \inf P_{s+t} f \leq (1/2) M(P_s)(\sup P_t g - \inf P_t g) \leq (1/2) M(P_s) M(P_t).$$

2° From Lemma 5.1 we know that P_t is a compact operator on $C(T)$. If $f \in \mathcal{B}_b(T)$ and $s > 0$ then $\{P_t f / t \geq s\}$ is equicontinuous, and hence $P_t f \to \mu(f)$ uniformly. On the other hand $K = \{P_s f / f \in \mathcal{B}(T), |f| \leq 1\}$ is a compact subset of $C(T)$ and the family of operators $\{P_t / t \geq s\}$ is equicontinuous. Then $P_t \to \mu \otimes 1$ uniformly on K, where the operator $\mu \otimes 1$ associates to each function $f \in C(T)$ the constant function $\mu(f)$. This implies the assertion.

Let T be a compact space with a countable base, L a local operator on T, and $\{P_t / t > 0\}$ a semigroup on $C(T)$ whose infinitesimal generator has $D(T,L)$ as domain and coincides with L as linear operator on this domain.

We suppose that $X_i = T$ and $L^i = a_i L$, $a_i \in R_+$, for each $i \in N$. Then $P_t^i = P_{a_i t}$ for any $t > 0$, $i \in N$. We also suppose that there exists a measure μ on T such that $P_t^x \to \mu \ (t \to \infty)$ for each $x \in T$.

6.5. Proposition

If the sequence $\{a_i / i \in N\}$ satisfies

(3) $$\sum_{i \in N} e^{-\theta a_i} < \infty \quad \text{for each} \quad \theta > 0 .$$

then $\sum_{i \in N} L_i$ satisfies (SFS).

Proof. From Lemma 6.4, 1° we deduce

$$M(P_t) < 2(M(P_r)/2)^{[t/r]}$$

for each $t, r > 0$, where $[t/r] \in N$ and satisfies $0 < t - [t/r] r < r$. Using 6.4, 2° we choose r such that $M(P_r) < 2$ and denote by $\theta = -\ln(M(P_r)/2)$. Then we deduce

$$M(P_{a_i t}) \leqslant 2e^{\theta} e^{-(\theta t/r)a_i}$$

Hence $\sum_{i \in N} M(P_t^i) < \infty$ and the proposition results from Theorem 6.3.

Chapter V

1. Local Operators on a l.c.a Group

Let X be a locally compact abelian group with a countable base and $\mathcal{H}$ a sheaf on X such that $(X,\mathcal{H})$ is a quasiharmonic space. We suppose that $\mathcal{H}$ is translation invariant, i.e. for any open set U and any $x\in X$, $f\in\mathcal{H}(U)$ if and only if $\tau_x f$ $\mathcal{H}(U+x)$, where $\tau_x f(x)=f(y-x)$. As a consequence if U is regular then $H^U f(y)=H^{U+x}(\tau_x f)(x+y)$ for any $y\in U$, $f\in C(\partial U)$ and $x\in X$.

1.1. Proposition. There exists an open set, U, and a C-strict potential, $p\in\mathcal{P}_b(U)\cap C(U)$, such that for any regular set U_o and any $x_o\in X$ we have

$$p-H^{U_o}p=\tau_{-x_o}(p-H^{U_o+x_o}p)$$

provided $\overline{U}_o$, $\overline{U}_o+x_o\subset U$.

Proof. Let V be an open set, $0\in V$ and $s\in\mathcal{P}_b(V)\cap C(V)$ a C-strict potential. We choose a relatively compact open set W such that $\overline{W}\subset V$, $0\in W$ and a function $\varphi\in C_c(W)$, $\varphi\geq 0$, $\varphi(0)>0$ and put $t=\varphi.s$. Then $t\in\mathcal{H}(V\setminus\overline{W})$ and there exists a D-regular set, U_1 , such that $0\in U_1$ and $t-H^{U_1}t$ is C-strict on U_1. Further we choose a regular set, U, such that $0\in U$, $\overline{U}\subset U_1$ and $U-U+W\subset V$. Then we put

$$p(y)=\int_{-W+U}(\tau_x t(y)-H^U(\tau_x t)(y))dx \quad \text{for} \quad y\in U.$$

Obviously $p\in\mathcal{P}_b(U)\cap C_o(U)$ and it is a C-strict potential. Now let U_o be a regular set, $0\in U_o$ and $x_o\in X$ such that $\overline{U}_o$, $\overline{U}_o+x_o\in U$. Then for $y\in U_o$ we have

$$(1)\qquad \tau_{-x_o}(p-H^{U_o+x_o}p)(y)=\int_{-W+U}(t(y+x_o-x)-H^U(\tau_x t)(y+x_o))dx-$$

$$-H^{U_o+x_o}\Big(\int_{-W+U}\tau_x t-H^U(\tau_x t))dx\Big)(y+x_o)=$$

$$=\int_{-W+U}(t(y+x_o-x)-H^{U_o+x_o}(\tau_x t)(y+x_o))dx \quad .$$

For fixed $y\in U_o$ we define $f:-W+U\to\mathbb{R}$, by $f(x)=t(y-x)-H^{U_o}(\tau_x t)(y)$.

Since $t \in \mathcal{H}(V\setminus\overline{W})$ we remark that supp $f \subset -W+U_o$. On the other hand $-W+x_o+U_o \subset -W+U$, and hence

$$\int_{-W+U} f(x)dx = \int_{-W+U} \tau_{x_o} f(x)dx$$

But this last term equals the last term in (1) and similar computations shows that

$$(p-H^{U_o}p)(y) = \int_{-W+U} f(x)dx \quad .$$

1.2. Proposition

Let V be an open set and $p_i \in P_b(V) \cap C(V)$, $p_i \neq 0$, $i=1,2$, such that for any regular set U and any $x \in E$

$$p_i - H^U p_i = \tau_{-x}(p_i - H^{U+x}p_i), \qquad i=1,2$$

provided $\overline{U}$, $\overline{U}+x \subset V$.

Then there exists $\alpha \in R_+$ such that $p_2 = \alpha p_1$

Proof. It is easy to see that for any regular U such that $\overline{U} \subset V$ we have $p_i > H^U p_i$ on U. Let $p=p_1+p_2$. A result of Mokobodzki [29] gives us a function $0 \leq f \leq 1$ such that $p_1 = f.p$. Let U be a regular set, $0 \in U$ and $U+U \in V$. We define $\varphi \in C_b(U)$ by $\varphi(x) = (1/\mu(U))\int_U f(x+y)dy$, where $\mu(U)$ is the Haar measure of U. One deduces

$$\varphi.(p-H^U p) = (1/\mu(U)\int_U (\tau_{-y}f).(p-H^U p)dy = p_1 - H^U p_1 \quad .$$

Now if U_o is regular and $\overline{U}_o + \overline{U}_o \subset U$, $x \in U_o$, then

$$\varphi.(p-H^{U_o}p) = p_1 - H^{U_o}p_1 = \tau_{-x}(p_1 - H^{U_o+x}p_1) =$$

$$= (\tau_{-x}\varphi).\tau_{-x}(p-H^{U_o+x}p) = (\tau_{-x}\varphi).(p-H^{U_o}p) \quad .$$

because $g.q = \tau_{-x_o}\tau(\tau_{x_o}g.\tau_{x_o}q)$ for any potential $q \in P_b(W)$, W open set, and $x_o \in X$ (use II.2.8). Further we get $\varphi = \tau_{-x}\varphi$ and deduce that φ is constant.

From the preceding propositions and Corollary II.6.5 we deduce:

1.3. Corollary. There exists a unique translation invariant local operator on X which is locally dissipative, possesses a base of P- and D-regular sets, and Ker $L=\mathcal{H}$.

1.4. Remark

The global resolvent $\{G_\lambda/\lambda>0\}$ and the global semigroup $\{P_t/t>0\}$ associated to such a local operator are translation invariant. Thus we have arrived at the situation studied by C.Berg and G.Forst [4]. For a P-harmonic group such a result was proved by J.Bliedtner [44].

2. Local Operators on Harmonic Spaces

First we present a result which makes clear the connection between the harmonic spaces in the meaning of C.Constantinescu and A.Cornea ([30] p.30) and continuous standard processes. It was firstly proved by J.C. Taylor [40] and(in a more general form)by J.Bliedtner and W.Hansen [5]. The short proof given by J.C. Taylor uses the following result of Ph. Courrège and P.Priouret [14] 2.1.7 Theorem:

2.1. Theorem. Let $(\Omega,\mathcal{M},\mathcal{M}_t,X_t,\theta_t,P^x)$ be a continuous standard process with state space E such that its transition function, $\{P_t/t>0\}$, and its resolvent, $\{R_\lambda|\lambda>0\}$, satisfy the following conditions:

(LC): $\lim_{t\to 0} N_t f=f$ uniformly on each compact set, for any $f\in C_c(E)$.

(SF): $R_\lambda f\in C_b(E)$, for each $\lambda>0$ and each function $f\in \mathcal{B}_b(E)$ such that $\{f\neq 0\}$ is relatively compact.

Then $P_{CU}f\in C(U)$ for each relatively compact open set, U, and each $f\in\mathcal{B}_b(E)$.

The next statement is more general than the result of J.C. Taylor and less general than the result of J.Bliedtner and W.Hansen. The proof retakes the method of J.C. Taylor.

2.2. Theorem. Let $(\Omega,\mathcal{M},\mathcal{M}_t,X_T,\theta_t,P^x)$ be a continuous standard process with state space E such that $P^x(T_{c\{x\}}<\zeta)>0$, for each $x\in E$. Assume also that the potential kernel, R_0 , $R_0f(x)=E^x[\int_0^\infty f(X_t)dt]$, is bounded and strong Feller, i.e. $R_0\mathcal{B}_b(E)\subset C_b(E)$.

Then the hyperharmonic sheaf, H^*, generated by the sweeping system $\{(P^x_{CU})x\in U|U$ relatively compact open set$\}$ defines a harmonic space (E,H^*). A function $f:V\to\overline{\mathbb{R}}_+$, (V open set) is excessive on V iff it is hyperharmonic on V. If $f<\infty$ then it is a natural potential on V iff f is a potential for the harmonic space (i.e. $f \in P(V)$). If U is a relatively compact open set, then the sweeping $(H^{U,x})_{x\in U}$ defined by the Perron-Wiener-Brelot solution to the Dirichlet problem coincides with $(P^x_{CU})_{x\in U}$.

<u>Proof</u>. First we note that the transition function, $\{P_t | t>0\}$ of the process satisfies condition (LC) of Theorem 1.1. This results from Proposition 2.5 of Chapter VI . Since the potential kernel satisfies $R_0(\mathcal{B}_b(E)) \subset C_b(E)$ we deduce that the resolvent of the process $\{R_\lambda\}$ satisfies condition (SF) of Theorem 1.1.

The axiom of convergence for H^* results from Theorem 1.1. Further let $x\in E$. Since $P^x(T_{C\{x\}}<\zeta)>0$, there exists a relatively compact open neighbourhood of x,V, such that $P^x(T_{CV}<\zeta)>0$. Then $P_{CV}1 \in H_{H^*}(V)$ and $P_{CV}1(x)>0$, which proves the axiom of positivity.

Now let U be a relatively compact open set. Since $G^U f(x)=:E^x[\int_0^{T_{CU}} f(X_t)dt]=R_0 f(x)-H^U R_0 f(x)$, $x \in U$, we deduce $G^U f \in C_b(V)$ for each $f\in\mathcal{B}_b(U)$. Hence each excessive function on U is lower semicontinuous and belongs to $H^*(U)$. Further let us suppose that there exist another open set V nad $h \in H_{H^*}(V)$ such that $\overline{U}\subset V$ and $h>0$. Let $s \in H^*(V)$. We choose $\alpha \in R$ such that $s+\alpha h\geq 0$ on a neighbourhood U_1 of $\overline{U}$. One deduces that $s+\alpha h$ is excessive on U_1 and $P_{CU}(s+\alpha h)(x) \leq s(x)+\alpha h(x)$, or $P_{CU}s(x)\leq s(x)$ for each $x\in U$. Let now $s \in H^*(V) \cap C(V)$, $s\geq 0$ and $t \in H^*(U)$ such that

$$\liminf_{\substack{x\to y\\ x\in U}} t(x)\geq s(y) \quad \text{for each} \quad y\in\partial U.$$

Then choosing a sequence of open sets, $\{U_n | n \in N\}$, such that $\overline{U}_n\subset U_{n+1}$ and $U=\bigcup_n U_n$, we compute:

$$t(x)\geq \liminf_{n\to\infty} P_{CU_n} t(x) \geq E^x[\liminf_{n\to\infty} t(X_{T_{CU_n}})] \geq$$

$$\geq E^x[s(X_{T_{CU}})]=P_{CU}s(x) \quad .$$

This and Hunt's balayage theorem show that $P_{CU}s$ is the solution of the Dirichlet problem in the sense of Perron-Wiener-Brelot. The same holds when s is of the form $s=Rg$, $g \in \mathcal{B}_b(X)$. If $f\in C_c(X)$ then $f=\lim_{n\to\infty} Gf_n$, uniformly on $\overline{U}$, where $f_n=n(f-nG_n f)$. Standard arguments (see [13] 2.4.2) shows that $H^U f$ is the solution of the Dirichlet problem in the sense of Perron-Wiener-Brelot. Thus U is resolutive. The reminder proof is obvious.

Now we sketch a short proof for the result of J.Bliedtner and W.Hansen from [5] :

1.3. Theorem. Let $(\Omega,M,M_t,X_t,\theta_t,P^x)$ be a continuous standard process with state space E such that $P^x(T_{C\{x\}}<\zeta)>0$ for each $x\in E$. Assume also that the potential kernel satisfies $R_o f(x)=E^x[\int_0^\infty f(X_t)dt]<\infty$ for each $x\in E$ and each $f\in C_c(E)$ and that each excessive function is the limit of an increasing sequence of continuous excessive functions.

Then the conclusion of Theorem 1.2 holds.

Proof. Let $\{K_n\}$ a sequence of compact sets such that $K_n\subset \overset{\circ}{K}_{n+1}$ and $E=\bigcup_n K_n$. Let $\{f_m^n\}$ be a family of continuous bounded excessive functions such that, for each $n\in N$, $f_m^n\leq f_{m+1}^n$ and $R_o\chi_{K_n}=\lim_m f_m^n$. Then each function f_m^n is a regular potential. We put

$$p=\sum_{n,m\in N}(1/\ n+m)(1/||f_m^n||)f_m^n .$$

p is a regular potential. From Theorem (3.13) p.167 in [6] we get a continuous additive functional A associated to p. If T is a stopping time such that $P^x(T>0)=1$ one easily see that $p(x)>P_T p(x)$, which shows that A is strictly increasing. Now we make a random time change with A (see(2.11) p.212 in [6]) and get a standard process which satisfies the requirements of Theorem 1.2. The family of all excessive functions is the same for both processes. Therefore Theorem 1.3 follows from Theorem 1.2.

Let X be a locally compact space with a countable base and $\mathcal{U}$ a hyperharmonic sheaf such that $1\in\mathcal{U}(X)$ and $(X,\mathcal{U})$ is a harmonic space.

2.3. Definition

Suppose that L is a local operator on X. We say that L is associated to $\mathcal{U}$ if there exists a family of open sets, $\{U_i/i\in N\}$, and a family of potentials, $\{p_i/i\in N\}$, such that

a) $\bigcup_{i\in N} U_i = X$,

b) $p_i \in C(U_i) \cap P(U_i)$, p_i is a strict potential in the meaning of [13] p.166, for each $i \in N$,

c) $p_i - p_j \in \mathcal{H}_{\mathcal{U}}(U_i \cap U_j)$, for each $i,j \in N$,

d) if $\varphi \in C_b(U_i)$, then $\varphi . p_i \in D(U_i, L)$ and $L(\varphi . p_i) = -\varphi$, for each $i \in N$,

e) $\mathcal{H}_{\mathcal{U}}(U) = \{f \in D(U,L) / Lf = 0\}$ for each open set $U \subset X$.

2.4. Proposition

1° Let $(X, \mathcal{U}')$ be another harmonic space. If there exists a local operator which is both associated to $\mathcal{U}$ and to $\mathcal{U}'$, then $\mathcal{U} = \mathcal{U}'$.

2° There exists (at least) a local operator on X which is associated to $\mathcal{U}$.

Proof. 1° It is trivial. 2° One can make a construction similar to that of Section 6. from Chapter II.

From now on we suppose that L is a local operator associated to $\mathcal{U}$ and U_i , p_i , $i \in N$ satisfy conditions a), b), c), d), e) of Definition 2.3.

2.3. Theorem. There exists a continuous standard process $(\Omega, \mathcal{M}, \mathcal{M}_t, X_t, \theta_t, P^x)$ with state space X such that for each $i \in N$,

(1) $$f.p_i(x) = E^x\left[\int_0^{T_{CU_i}} f(X_t)\,dt\right], \quad f \in C_b(U_i), \quad x \in U_i .$$

If another continuous standard process satisfies relation (1), then it has the same transition function. Furthermore the family of all excessive functions on an open set U coincides with $\mathcal{U}_+(U)$.

Proof. Let $i \in N$ be fixed. From Theorem 10.2.1 and Proposition 10.2.3 of C.Constantinescu and A.Cornea [13] and from Theorem 3 of W. Hansen [19] p.208 we get a standard process $(\Omega, \mathcal{M}, \mathcal{M}_t, X_t, \theta_t, P^x)$ with state space U_i such that

$$f.p_i(x) = E^x\left[\int_0^{\infty} f(X_t)\,dt\right], \quad f \in C_b(U_i), \quad x \in U_i$$

and the family of all excessive functions coincides with $\mathcal{U}_+(U_i)$. If U is a relatively compact open set, $\overline{U} \subset U_i$, then by using Hunt's balayage theorem one deduces $P^x_{CU}=H^{U}{}^{x}$ for $x\in U$. Further Theorem I.2.4 shows that the process is continuous. We also deduce, for $x\in U$ and $f\in C_b(U_i)$,

$$E^x\left[\int_0^{T_{CU}} f(X_t)\,dt\right]=f.p_i(x)-H^U(f.p_i)(x)=f.(p_i-H^U p_i)(x)$$

Now for $j\neq i$ such that $\overline{U}\subset U_j$ we have $p_j-H^U p_j=p_i-H^U p_i$. We deduce that we get same transition function on U if either the process on U_i associated to p_i or the process on U_j associated to p_j is used. The reminder proof is straightforward.

2.6. Proposition

For each $\lambda>0$ there exists a hyperharmonic sheaf, $\mathcal{U}_\lambda$, such that $(X,\mathcal{U}_\lambda)$ is a harmonic space and L_λ is associated to $\mathcal{U}_\lambda$. If U is an open set, then $\mathcal{U}_{\lambda+}(U)$ coincides with the family of all λ-excessive functions on U with respect to the process constructed in Proposition 2.5.

Proof. Let $(\Omega,M,M_t,X_t,\theta_t,P^x)$ be the process from Proposition 2.5. If U is an open set we put the notation

$$G^U_\lambda f(x)=E^x\left[\int_0^{T_{CU}}\exp(-\lambda t)f(X_t)\,dt\right],\ \text{for } \lambda\geq 0,\ f\in \mathcal{B}_b(U) \text{ and } x\in U.$$

Then for a relatively compact open set, U, with $\overline{U}\subset U_i$ for some $i\in N$, we have $G^U_o f=G^{U_i}_o f-H^U G^{U_i}_o f$, and hence $G^U_o f\in D(U,L)$, $LG^U_o f=-f$ on U for $f\in C_b(U_i)$. Further we deduce $G^U_\lambda f\in D(U,L)$ and $L_\lambda G^U_\lambda f=-f$ for each $\lambda\geq 0$ and $f\in C(\overline{U})$, on account of relation $G^U_\lambda=G^U_o-\lambda G^U_o G^U_\lambda$.

Now we note that L is locally closed: let $\{\varphi_n\}\subset D(V,L)$, V open, $V\subset U_i$ for some $i\in N$, $\varphi_n\to\varphi$ and $L\varphi_n\to\psi$, uniformly on each compact subset of V. If U is a relatively compact open set such that $\overline{U}\subset V$, then $\varphi_n+G^U_o L\varphi_n\in \mathcal{H}_{\mathcal{U}}(U)$ and letting $n\to\infty$ we get $\varphi+G^U_o\psi\in\mathcal{H}_{\mathcal{U}}(U)$, and hence $\varphi\in D(U,L)$, $L\varphi=\psi$.

If U is relatively compact and $\overline{U}\subset U_i$ for some $i\in N$, we have $P^\lambda_{CU}G^{U_i}_\lambda f=G^{U_i}_\lambda f-G^U_\lambda f$ for $f\in C_b(U_i)$, and hence $P^\lambda_{CU}G^{U_i}_\lambda f\in D(U,L)$,

$L_\lambda P^\lambda_{CU} G^{U_i}_\lambda f=0$. Further one makes an approximation and deduces $P^\lambda_{CU} f \in D(U,L)$, $L_\lambda P^\lambda_{CU} f=0$ for each $f \in C(\partial U)$. Conversely let $f \in D(V,L)$, V open set $\overline{U} \subset V$, be such that $L_\lambda f=0$. We choose another relatively compact open set, U', such that $\overline{U} \subset U'$, $\overline{U'} \subset V$. Then we have

$$f+\lambda G_o^{U'} f \in \mathcal{H}_{\mathcal{U}}(U') ,$$

$$f+\lambda G_o^{U'} f = P_{CU} f + \lambda P_{CU} G_o^{U'} f \quad \text{or} \quad f = P_{CU} f - \lambda G_o^U f .$$

One deduces $f=\sum_{k=0}^{n} (-\lambda G_o^U)^k P_{CU} f + (-\lambda G_o^U)^{n+1} f$. For small enough U we have $||G_o^U 1|| \leq 1/\lambda$. It follows

$$f=\sum_{k=0}^{\infty} (-\lambda G_o^U)^k P_{CU} f .$$

On the other hand a straightforward computation proves

$$(G_o^U)^k P_{CU} f(x) = (1/k!) E^x [(T_{CU})^k f(X_{T_{CU}})] ,$$

which shows $f(x)=P^\lambda_{CU} f(x)$ for each $x \in U$.
Now we take $\mathcal{U}_\lambda$ to be the hyperharmonic sheaf defined by the sweeping system $\{(P^{\lambda,x}_{CU})_{x\in U} | U$ relatively compact open set$\}$ and using Theorem 2.2 deduce the statement of Proposition 2.6.

2.7. Theorem. Let $\{P_t \ t>0\}$ be the transition function of the process constructed in Theorem 2.4 and $G_\lambda = \int_0^\infty \exp(-\lambda t) P_t dt$, $\lambda>0$. Then for each $\lambda>0$,

1° $G_\lambda f \in D(X,L)$ and $L_\lambda G_\lambda f=-f$ if $f \in C_b(X)$,

2° for each open set U, $f \in C_c(U)$ and $\varepsilon>0$ there exists $g \in C_c(U)$ such that $|f-G_\lambda g|<\varepsilon$.

3° $P_t(C_o(E)) \subset C_b(E)$.

The proof is quite similar to the proof of Theorem III.1.1.

In order to state a result simillar to Theorem IV.5.3 we remark that L has a property similar to (SFR). Also the property (SFS), the operator $L-d/dt$ and the sum of two local operators associated to harmonic spaces can be defined as in Chapter IV.

2.7. Theorem. The following properties are equivalent:

1° L satisfies (SFS).

2° L-d/dt is associated to a harmonic space.

3° If $(X^o, \mathcal{U}^o)$ is a harmonic space such that $1 \in \mathcal{U}^o(X^o)$, X^o has a countable base, and L^o is a local operator associated to $\mathcal{U}^o$, then $L+L^o$ is associated to a harmonic space on $X \times X^o$.

4° If $X^o = T \times R$, L^o is the local closure of $\partial^2/\partial x^2 - \partial/\partial t$, and $\mathcal{U}^o$ is the hyperharmonic sheaf associated to L^o, then $L+L^o$ is associated to a harmonic space.

5° If U is an open set, $U \subset X$, and $\{P_t / t>0\}$ is a sub-Markov semigroup of kernels which is also a (C_o)-class semigroup of operators on a Banach space $F \subset C_b(U)$ whose infinitesimal generator, Δ, has a domain, $D(\Delta)$, such that $D(\Delta) \subset D(U,L)$, $\Delta = L$ as linear operators on $D(\Delta)$, and $C_o(U) \subset \overline{D(\Delta)} = F$, then P_t is strong Feller for each $t>0$.

The proof of this theorem is quite simillar to the proof of Theorem IV.5.3.

VI. FELLER RESOLVENTS

1. Convex Cones of Lower Semicontinuous Functions

In this section we shall prove an improvement of a wellknown result (see for example [30] Proposition 1 p.226). The proof follows from the original idea and an idea of C.Constantinescu and A.Cornea [13] (Lemma from page 160).

Let C be a convex cone of lower semicontinuous nonnegative functions on a locally compact space, E, which has a countable base. Assume that for each $x \in E$, there exists a function $c \in C$ such that $0 < c(x) < \infty$. Let us denote by C^* the family of all numerical nonnegative universally measurable functions, f, such that $\mu(f) < f(x)$ for each $x \in E$ and each measure μ that satisfies $\mu(c) < c(x)$ for any $c \in C$.

Let $f: E \longrightarrow R$ be a function such that there exists $c_0 \in C$ with $f \leq c_0$. We shall use the notation

$$Rf = \inf \{c \in C^* \mid f \leq c\}$$

It follows that $Rf \leq 0$ if the function f satisfies $f \leq 0$. We denote by D the family of all functions $f \in C(E)$ which have the following properties:

1° there exists $c \in C$ such that $|f| \leq c$

2° $\inf \{R(|f| \chi_{CK}) \mid K \text{ compact set}\} = 0$.

Obviously D is a vector lattice that contains $C_c(E)$ and $Rf < \infty$ for each $f \in D$.

1.1. Theorem

Let f be an upper semicontinuous function such that there is $g \in D$ with $f \leq g$. Then for each $x \in E$ there exists a nonnegative measure μ such that

a) $\mu(c) \leq c(x)$ for each $c \in C$,

b) $Rf(x) = \mu(f)$.

Proof

We define, for each $g \in D$, $p(g) = Rg(x)$. One easily verifies that p is sub-liniar on D.

Let $\mu:D\longrightarrow R$ be a linear functional such that $\mu(g)\leq p(g)$ for each $g\in D$. Then for $g\leq 0$ we have $\mu(g)\leq p(g)=0$ and hence μ is nonnegative. The restriction $\mu|_{C_c(E)}$ define a nonnegative measure on E, which we shall denote by $\bar{\mu}$. Now let $g\in D$, $g\geq 0$ and choose a sequence $\{h_n\}\subset C_c(E)$ such that $0\leq h_n\leq h_{n+1}\leq 1$ and $\bigcup_n \overbrace{\{h_n=1\}}^{\circ}=E$. Then $gh_n, g(1-h_n)\in D$ and $R(g(1-h_n))\ \to 0$, as $n\to\infty$. Therefore $\mu(g(1-h_n))\leq R(g(1-h_n))(x)\to 0$, and hence

$$\bar{\mu}(g)=\lim_{n\to\infty}\bar{\mu}(gh_n)=\lim_{n\to\infty}\mu(gh_n)=\mu(g)-\lim_{n\to\infty}\mu(g(1-h_n))=\mu(g)$$

We conclude that $\bar{\mu}=\mu$ on D. On the other hand for $c\in C$, let $g\in C_c(E)$ be such that $g\leq c$. Then

$$\mu(g)\leq p(g)\leq c(x)\quad ,$$

which leads to $\mu(c)\leq c(x)$.

Conversely let μ be a nonnegative measure on E such that $\mu(c)\leq c(x)$ for each $c\in C$. Then μ is finite on D and $\mu(g)\leq p(g)$ for each $g\in D$.

Now let us suppose that $f\in D$. Then the assertion of the theorem results from the Hahn-Banach theorem applied on the space D.

If f is upper semicontinuous, let us consider a sequence $\{f_n\}\subset D$ such that $f_{n+1}\leq f_n$ and $f=\inf_n f_n$. The set $B=\{\mu\in D'\,|\,\mu(g)\leq p(g)$ for each $g\in D\}$ is a compact set in the topology $\sigma(D',D)$, because $-p(-g)\leq\mu(g)\leq p(g)$ for each $g\in D$ and each $\mu\in B$. The functions $\bar{f},\bar{f}_n:B\to R$ defined by $\bar{f}(\mu)=\mu(f)$, $\bar{f}_n(\mu)=\mu(f)$ for $\mu\in B$ satisfy $\bar{f}=\inf\bar{f}_n$ and $\bar{f}_n$, $n\in N$ are continuous on B. Therefore from Lemma 1.2 stated below it follows

$$\sup_B \bar{f}=\inf_n\left[\sup_B \bar{f}_n\right]$$

From the first part of the proof we know $Rf_n(x)=\sup_B \bar{f}_n$. Since $Rf(x)\leq Rf_n(x)$, for each $n\in N$, and $\bar{f}(\mu)=\mu(f)\leq Rf(x)$ for each $\mu\in B$ we get

$$\sup_B \bar{f}\leq Rf(x)\leq\inf_n\left[\sup_B \bar{f}_n\right]$$

Since $\bar{f}$ is an upper semicontinuous function on a compact space there exists $\mu_o \in B$ such that $\mu_o(f) = \sup_B \bar{f}$.

1.2. Lemma

Let K be a compact space and (f_n) a lower bounded decreasing sequence of upper semicontinuous numerical functions. Then the following equality holds:

$$\sup_{x \in K} (\inf_n f_n(x)) = \inf_n (\sup_{x \in K} f_n(x)) .$$

2. Feller Resolvents

Let E be a locally compact space with a countable base. In this section we shall study a sub-Markov resolvent of kernels $\{V_\lambda/\lambda>0\}$ which has the following property of W.Feller:

$$V_\lambda C_b(E) \subset C_b(E) \ , \text{ for each } \quad \lambda \geq 0 \ .$$

We also assume that for each $f \in C_c(E)$ and each $x \in E$,

(1) $$\lim_{\lambda\to\infty} \lambda V_\lambda f(x) = f(x) \ .$$

2.1. Remark. The following fact is wellknown (see [11]): let $\{V_\lambda | \lambda>0\}$ be a sub-Markov resolvent of kernels that satisfies property (1). If f is a lower semicontinuous nonnegative function such that $\lambda V_\lambda f \leq f$ for each $\lambda>0$, then f is excessive. Endeed if $g \in C_c(E)$ is such that $g \leq f$ then $g = \lim_{\lambda\to\infty} \lambda V_\lambda g \leq \lim_{\lambda\to\infty} \lambda V_\lambda f \leq f$. Since $f = \sup\{g \in C_c(E) | g \leq f\}$ it follows $f = \lim_{\lambda\to\infty} \lambda V_\lambda f$.

We shall denote by S the family of all excessive functions on E and by S_c the subfamily of all continuous excessive functions. For each bounded function f we define

$$Rf = \inf\{g \in S | f \leq g\}$$

$$^cRf = \inf\{g \in S_c | f \leq g\}$$

Obviously $Rf \leq {}^cRf$. G.Mokobodzki proved in [30] (see p.220-221) that the cone of all excessive functions associated to an arbitrary sub-Markov resolvent is a potential cone. In our situation this property is stated in the following theorem.

2.1'. Theorem

If $s,t \in S$, then $R(s-t) \in S$ and $s-R(s-t) \in S$.

The following three results stated in the theorem from below are easy consequences of some results of G.Mokobodzki [30]. (See Theorem 6 p.212, p.221, Proposition 8, p.229, Theorem 12, p.236, Proposition

14, p.232 and Proposition 16, p.233 in [30]. The proof of 2° results by using Theorem 1.1 instead of Proposition 1 from p.226 in [30]).

2.2. <u>Theorem</u>

1° If f is a bounded lower semicontinuous function, then Rf is also lower semicontinuous.

2° If g is an upper semicontinuous function and there exists a bounded continuous function f such that $g \leq f$ and

$$\inf \{R(f\chi_{CK}) \mid K \text{ compact set} \subset E\} = 0 , \tag{2}$$

then $Rg = {}^{c}Rg$. Particularly Rg is upper semicontinuous.

3° If f is a bounded continuous function which fulfils relation (2), then Rf is a continuous function.

2.3. <u>Remark</u>. 1° If f is a bounded lower semicontinuous function, then from the above theorem and Remark 2.1 it follows that Rf is excessive.

2° If f is a continuous function with compact support, then condition 3° of the above theorem is obviously fulfilled. Therefore one deduces that each lower semicontinuous excessive function is the limit of an increasing sequence of bounded continuous excessive functions.

3° Let f be a bounded continuous excessive function which fulfils relation (2). Then f fulfils the following condition:

$$\inf \{{}^{c}R(f\chi_{CK}) \mid K \text{ compact set}\} = 0$$

Endeed, let $\{g_n\}$ be a sequence in $C_c(E)$ such that $0 \leq g_n \leq g_{n+1} \leq 1$ and $\bigcup_n \overset{\circ}{\overbrace{\{g_n=1\}}} = E$. Then $R(f(1-g_n))$, $n \in N$ are continuous and

$${}^{c}R(f\chi_{\{g_n=0\}}) \leq R(f(1-g_n)) \leq R(f\chi_{\{g_n<1\}})$$

Since each compact set K satisfies $K \subset \overset{\circ}{\overbrace{\{g_n=1\}}}$ for some $n \in N$ we deduce

$R(f\chi_{\{g_n<1\}}) \to 0$, which implies the assertion.

4° Now let $(\Omega, M, M_t, X_t, \theta_t, P^x)$ be a standard process with state space E. Since $E^x[f(X_t)] \to f(x)$, $(t \to 0)$ for each $x \in E$ and each $f \in C_c(E)$, it follows that the resolvent of the process satisfies relation (1). If the potential kernel of the process is finite, i.e.

$$G1(x) = E^x[\zeta] < \infty \quad , \quad \text{for each } x \in E,$$

then from Hunt's theorem (see [6] p.141) for each compact set K and each $x \in K$, it follows

$$\inf \{s(x) \mid G1 < s \text{ on } CK, \text{ s excessive}\} = E^x[\zeta - T_{CK}] .$$

If K_n is an increasing sequence of compact sets such that $K_n \subset \mathring{K}_{n+1}$, $T_{CK_n} \to \zeta$, and hence

$$\inf_n \inf \{s(k) \mid G1 \leq s \text{ on } CK_n, \text{ s excessive}\} = 0$$

In the sequell we want to associate a Hunt process to the given resolvent (V_λ). Therefore <u>from now on we assume that $V_o 1$ satisfies relation (2)</u>.

The family of all excessive functions that satisfy relation (2) will be denoted by P. Our assumption implies $V_o C_{b+}(E) \subset P$. We put

$$T = \{f \in C_c(E) \mid \text{there exist } s,t \in P \cap C_b(E) \text{ such that } f = s-t\}$$

Obviously T is a vector lattice. We assert that T linearly separates the points of E, i.e. for each $x,y \in E$ there exist $f,g \in T$ such that

$$f(x)g(y) \neq f(y)g(x) .$$

Since $C_c(E)$ has this property from condition (1) and the relation $V_\lambda f = V(f - \lambda V_\lambda f)$ we first deduce that $V(C_b(E))$ linearly separates the points of E. Then $P \cap C_b(E)$ has the same property, because $V(C_{b+}(E)) \subset P \cap C_b(E)$.

Now let f $P \cap C_b(E)$. If $g \in S_c$ then $\min(f,g) \in P \cap C_b(E)$. If $f \leq g$

on CK for some compact set K then $f-\min(f,g)\in T$, and the assertion follows on account of Remark 2.3.3°.

The Stone-Weierstrass theorem implies $T=C_0(E)$. Further Theorem 3.4 of J.C. Taylor [45] implies the following result:

2.4. Theorem

There exists a standard process $(\Omega, F, F_t, X_t, \theta_t, P^x)$ with state space E such that for each $x\in E$, $\lambda\geq 0$ and $f\in B_b(E)$,

$$E^x\left[\int_0^\infty \exp(-\lambda t) f(X_t)\,dt\right]=V_\lambda f(x)$$

Let us denote by $\{P_t\}$ the transition function of the process given by the above theorem.

2.5. Proposition

For each $f\in C_0(E)$, $\lim_{t\to 0} P_t f=f$ uniform on each compact set.

Proof

Let $f\in P\cap C_b(E)$. The sequence $f_n=nV_n f$ is increasing and $\lim f_n=f$. Dini's theorem implies that the convergence is uniform on each compact set. Further since

$$P_t f_n=P_t V(f-nV_n f)=\int_t^\infty P_t(f-nV_n f)\,dt$$

we get $f_n-P_t f_n\leq t\|f-nV_n f\|$, and hence $P_t f_n \to f_n$, uniform. The inequality $P_t f_n\leq P_t f\leq f$ shows that $P_t f \to f$ uniform on each compact set. Then the same holds for each $f\in T$, and since T is dense in $C_0(E)$ the proposition follows.

In order to show that the semigroup of the process given by Theorem 2.4 is in fact a Hunt semigroup we first give the next two lemmas.

2.6. Lemma

Let $(\Omega,\mathcal{M},\mathcal{M}_t,Y_t,\theta_t,P^x)$ be a standard process with state space E. Let Δ be the Alexandrov point if E is noncompact or an additional isolated point if E is compact and set $E_\Delta=E\cup\{\Delta\}$. Assume that for each pair $x,y\in E_\Delta$, $x\neq\Delta$ there exist two finite excessive functions s,t such that $s-t\geq 1$ on a neighbourhood of x and $s-t\leq 0$ on a neighbourhood of y. Then $\lim\limits_{\substack{t\to\zeta\\ t<\zeta}} Y_t$ exists in E_Δ a.s.

Proof

Let U_1 , U_2 be open sets in E_Δ and s,t finite excessive functions such that $s-t\geq 1$ on $\overline{U}_1$ and $s-t\leq 0$ on U_2. We are going to prove that the set

(4) $$M=\{\omega\in\Omega/\text{there exists two sequences } (t_n),(t'_n) \text{ such that } t_n\to\zeta(\omega),\ t'_n\to\zeta(\omega),\ Y_{t_n}(\omega)\in U_1,\ Y_{t'_n}(\omega)\in U_2\}$$

is negligible.

Let us define $T_1=T_{U1}$ and $T_{k+1}=T_k+T_{Ui}\circ\theta_{T_k}$, where i is taken such that i=1 if k is even and i=2 if k is odd. Then

$$M=\bigcap_{n\geq 1}\{T_n<T_{n+1}\} \quad .$$

Since $Y_{T_{2k+1}}\in U_1$ and $Y_{T_{2k}}\in U_2$ on $\{T_{2k}<T_{2k+1}<\zeta\}$ we deduce

$$P^x(\{T_{2k}<T_{2k+1}<\zeta\})\leq E^x[(s-t)(Y_{T_{2k+1}})-(s-t)(Y_{T_{2k}})] \quad .$$

Further we have

$$nP^x(M)\leq \sum_{k=1}^{n} E^x[(s-t)(Y_{T_{2k+1}})-(s-t)(Y_{T_{2k}})]\leq$$

$$\leq \sum_{k=1}^{n} E^x[t(Y_{T_{2k+1}})-t(Y_{T_{2k}})]\leq E^x[t(Y_{T_3})]\leq t(x) \quad ,$$

because $\{s(Y_{T_n})\},\{t(Y_{T_n})\}$ are supermartingales. Thus we deduce $P^x(M)=0$ for each $x\in E$.

The condition from the statement allows as to choose a

countable family $\{(U_1^n, U_2^n)\}$ of pairs of open sets in E_Δ and a family $\{(s^n,t^n)\}$ such that s^n, t^n, $n\in N$ are finite excessive functions $s^n-t^n\geq 1$ on U_1^n and $s^n-t^n\leq 0$ on U_2^n and for each pair $(x,y)\in E\times E_\Delta$ there exists $n\in N$ such that $x\in U_1$, $y\in U_2$. Then denoting by M_n the set defined by (4) for (U_1^n, U_2^n) we have

$$\{\omega\in\Omega \mid \lim_{\substack{t\to\zeta(\omega)\\ t<\zeta(\omega)}} Y_t(\omega) \text{ do not exists in } E_\Delta\}\subset \bigcup_n M_n .$$

and the desired conclusion follows.

2.7. Lemma

Let $(\Omega,F,F_t,Y_t,\theta_t,P^x)$ be a standard process with state space E such that $\lim_{\substack{t\to\zeta\\ t<\zeta}} Y_t$ exists in E_Δ a.s. Let $\{G_\lambda \mid \lambda>0\}$ be its resolvent and assume that for each $f\in C_c(E)$, $\lim_{\lambda\to\infty}\lambda G_\lambda f=f$ uniform on each compact subset of E. Then the process is a Hunt process.

Note. In this lemma and in the next corollary F and F_t denote the canonical σ-fields associated to a Markov process.

Proof.

Let $\{T_n\}$ be an increasing sequence of stopping times and $T=\lim_{n\to\infty} T_n$. Let $L=\lim_{n\to\infty} Y_{T_n}$ and put

$$M=\{T=\zeta<\infty \quad\text{and}\quad L\in E\}$$

We are going to prove that M is negligible. First we note that for each bounded nonnegative universally measurable function f, $\{e^{-t}V_1f(X_t)\}$ is a nonnegative supermartingale and

$$E^x[e^{-T_n}V_1f(X_{T_n})]=E^x[e^{-T_n}\int_{T_n}^{\infty} f(X_t)dt] \to E^x[e^{-T}\int_T^{\infty}(X_t)dt] .$$

Therefore $\lim_{n\to\infty} e^{-T_n}V_1f(X_{T_n})=e^{-T}V_1f(X_T)$ a.s.

Now let $f \in C_c(E)$. From the relation $V_k f = V_1(f-(k-1)V_k f)$ we deduce that $f_k = kV_k f$ satisfies

$$\lim_{n\to\infty} e^{-T_n} f_k(Y_{T_n}) = e^{-T} f_k(Y_T) \quad \text{a.s.},$$

and hence $\lim_{n\to\infty} f_k(Y_{T_n})=0$ a.s. on M. On the other hand for each $\omega \in M$, the set $\{Y_{T_n}(\omega) \mid n\in N\} \cup \{L(\omega)\}$ is compact. Since $f_k \to f$ uniform on each compact set we deduce $\lim_{n\to\infty} f(Y_{T_n})=0$, a.s. on M. But $\lim_{n\to\infty} f(Y_{T_n})=f(L)$ a.s. on account of the continuity of f, which implies $f(L)=0$ a.s. Since f is arbitrary choosen we conclude that M is negligible.

2.8. Corollary

The process $(\Omega, F, F_t, X_t, \theta_t, P^x)$ given by Theorem 2.4 is a Hunt process.

Proof

For each $f \in T$ we have $\lim_{\lambda\to\infty} \lambda V_\lambda f = f$ uniform on each compact set. Since T is dense in $C_0(E)$ we deduce $\lim_{\lambda\to\infty} \lambda V_\lambda f = f$ uniform un each compact set for each $f \in C_0(E)$. The corollary follows from the preceding two lemmas.

3. Excessive Functions for Feller Resolvents

Let $(\Omega, M, M_t, X_t, \theta_t, P^x)$ be a standard process with state space E and suppose that its resolvent $\{V_\lambda | \lambda>0\}$ has the following property: $V_\lambda C_b(E) \subset C_b(E)$, for each $\lambda>0$. In this section we are going to prove a criterion of excessiveness. The proof makes use of the Choquet boundary associated to a convex cone of lower semicontinuous functions on a compact topological space. Namely we use Bauer's minimum principle.

3.1. Theorem

Let f be a bounded continuous nonnegative function on E. Assume that for each $x \in E$ there exists a base of neighbourhoods of $x, \mathcal{U}(x)$, such that

$$P_{CW} f(x) \leq f(x) \quad \text{for each} \quad W \in \mathcal{U}(x).$$

Then f is an excessive function.

Proof

In order to simplify the exposition we first assume that the potential kernel V_o has also the property $V_o C_b(E) \subset C_b(E)$. From Remark 2.3, 4° one deduces that all results of Section 2.apply for our resolvent. Let us denote by $g = f - \lambda V_\lambda f$, $\varphi = \max(0, g)$, $\psi = \max(0, -g)$. Then $V_\lambda f = V_o(f - \lambda V_\lambda f) = V_o \varphi - V_o(\psi)$. From Remark 2.3, 3° we know that

$$\inf \{t \in C(E) \,|\, t \text{ is excessive and } V_o \varphi \leq t \text{ on } CK, \text{ for some compact set } K\} = 0 .$$

Therefore, in order to show $g \geq 0$, it suffices to show $g + t \geq 0$, for each continuous excessive function t such that $V_o \varphi \leq t$ on CK for some compact set K. For such a function t let us suppose that $\alpha = \inf (t+g) < 0$. Then $K_o = \{x \in E / (t+g)(x) = \alpha\}$ is a compact set because K_o must satisfy $K_o \subset K$. Also K_o must satisfy $K_o \subset \{g<0\}$.

Now let $x \in K_o$. Choose $W \in \mathcal{U}(x)$ such that $W \subset \{g<0\}$. Since $\{g<0\} \cap \{\varphi>0\} = \emptyset$ we have

$$E^x[\int_0^{T_{CW}} (x_t)dt]=0, \quad \text{which implies}$$

$$P_{CW}(V_o\varphi)(x)=E^x[\int_{T_{CW}}^{\infty} \varphi(X_t)dt]=V_o\varphi(x) \quad .$$

Since $g=f-\lambda V_o\varphi+\lambda V_o\psi$ we deduce $P_{CW}g(x)\leq g(x)$. Further on account of $\alpha\leq t+g$ and $P_{CW}(1)(x)\leq 1$ we get $\alpha\leq P_{CW}(\alpha)(x)\leq P_{CW}(t+g)(x)\leq t(x)+g(x)=\alpha$. It follows $P_{CW}(t+g-\alpha)(x)=0$, which shows $P_{CW}(\chi_{E\setminus K_o})=0$. If we denote by μ_x the measure on K_o defined by $\mu_x(f)=P_{CW}(f)(x)$ we see that $\mu_x(1)=1$, $\mu_x(s)<s(x)$ for each excessive function s, $\mu_x(t+g)\leq(t+g)(x)$ and $\mu_x(\overset{o}{w})=0$ because $X_{T_{CW}} \in \overline{CW}$, P^x-a.s.

Now we can apply Lemma 1.5 of Chapter II for the space K_o the cone of all lower semicontinuous excessive functions and the function g+t. We get $g+t\geq 0$. Finally we conclude $f\leq\lambda V_\lambda f$ and from Remark 2.1 deduce that f is excessive.

Now let us treat the general case (where we allow the potential kernel to be nonfinite). For $\lambda>0$ we first deduce $P^\lambda_{CW}f(x)\leq P_{CW}f(x)\leq f(x)$ for any $W\in\mathcal{U}(x)$ and any $x\in E$. Then from the first part of the proof we deduce that f is λ-excessive. Since λ is arbitrary it follows that f is excessive.

The above theorem can be stated in the following more general form:

3.2. Theorem

Let f be a continuous bounded function on E such that $\inf\{R(-f\chi_{CK}) \mid K \text{ compact set} \subset E\}=0$, where

$$R(-f\chi_{CK})=\inf\{t \mid t \text{ is excessive and } -f\chi_{CK}\leq t\}$$

Assume that for each $x\in E$ there exists a base $\mathcal{U}(x)$ of neighbourhoods of x such that

$$P_{CW}f(x)\leq f(x) \text{ for each } W\in\mathcal{U}(x) \quad .$$

Then f is nonnegative and excessive.

Proof

Let K be a compact set. From Theorem 2.2, 1° we know that $t=R(-f\chi_{CK})$ is lower semicontinuous. Let us suppose that $\inf (t+f)=\alpha<0$. Then put $K_o=\{x\in E \mid (t+f)(x)=\alpha\}$. It follows that K_o is a compact subset of K. Further we apply Lemma 1.5 of Chapter II and deduce $t+f\geq 0$ just like in the preceding proof. The assumption from the statement implies $f\geq 0$. The theorem results from the preceding one.

3.3. Remark. T.Watanabe in [41] proved other excessiveness criteria for a resolvent satisfying the condition $V_\lambda C_b(E)\subset C_b(E)$, for $\lambda>0$. Our results do not follow from his because we let the family $\mathcal{U}(x)$ to depend on x.

4. The Local Character

Let E be a locally compact space with a countable base and $(V_\lambda | \lambda \geq 0)$ a sub-Markov resolvent of kernels on E that satisfies the conditions assumed in Section 2. We shall use the notation from Section 2. Particularly $(\Omega, M, M_t, X_t, \theta_t, P^x)$ will be a Hunt process such that for each $f \in C_b(E)$, $\lambda \geq 0$, $x \in E$ the following relation holds:

$$V_\lambda f(x) = E^x[\int_0^\infty \exp(-\lambda t) f(X_t) dt] \quad .$$

In this section we shall characterise those resolvents (V_λ) which are associated to continuous Markov processes. In the sequel we shall use the following consequence of a result of G.Mokobodzki.

4.1. Theorem

Let $t \in P \cap C(E)$. There exists a unique kernel, G_t, on E such that $G_t 1 = t$ and for each $f \in C_{b+}(E)$,

$$G_t f \in P \cap C(E) \quad \text{and}$$

$$R(\chi_A G_t f) = G_t f \ , \quad \text{where } A = \text{supp } f.$$

The proof follows from Theorem 3 of Ch.IV in [30] and the next lemma.

4.2. Lemma

Let t be in $P \cap C(E)$. Then there exists a sequence $\{t_n\}$ in $P \cap C(E)$ such that $t = \sum_n t_n$ and for each $n \in N$ there exists a compact set K_n such that $R(t_n \chi_{K_n}) = t_n$.

Proof

Let $\{g_n\}$ be a sequence in $C_c(E)$ such that $0 \leq g_n \leq g_{n+1}$ and $\bigcup_n \overset{\circ}{\overline{\{g_n = 1\}}} = E$. We define $t_0 = 0$ and

$$t_{n+1} = R(t - \sum_{k \leq n} t_k - R((t - \sum_{k \leq n} t_k)(1 - g_{n+1}))) \ .$$

Next we are going to prove by induction that the sequence $\{t_n\}$ has the following properties:

$$t_n \in \mathcal{P} \cap C(E) \ , \qquad t - \sum_{k\leq n} t_k \in \mathcal{P} \cap C(E) \ .$$

Suppose that the above properties are true. Let us prove them with n+1 instead of n. First we note that

$$R((t-\sum_{k\leq n} t_k)(1-g_{n+1})) = t - \sum_{k\leq n} t_k \quad \text{on} \quad \{g_{n+1}=0\} \ .$$

From Theorem 2.2, 3° and Remark 2.3, 1° it follows that $R((t-\sum_{k\leq n} t_k)(1-g_{n+1}))$ is a continuous excessive function. Then the same arguments imply that t_{n+1} is also a continuous excessive function. From Theorem 2.1′ it follows that $t-\sum_{k\leq n} t_k - t_{n+1}$ is also excessive.

Further the inequality $t_{n+1} \leq t$ implies $t_{n+1} \in \mathcal{P}$, and similarly we deduce $t - \sum_{k\leq n+1} t_k \in \mathcal{P}$.

Now for each $n \in N$ we put $K_n = \text{supp } g_n$ and remark that $t_{n+1} = R(t_{n+1}\chi_{K_{n+1}})$.

From the definition of t_{n+1} we deduce

$$t - \sum_{k\leq n} t_k - t_{n+1} \leq R((t-\sum_{k\leq n} t_k)(1-g_{n+1})) \quad .$$

Further $t - \sum_{k\leq n+1} t_k \leq R(t(1-g_{n+1})) \leq R(t\chi_{CK_{n+1}})$.

Since the last term tends to zero we get $t = \sum_{k=1}^{\infty} t_k$.

4.3. Notation

If $t \in \mathcal{P} \cap C_b(E)$ and $f \in \mathcal{B}_b(E)$ we shall use the notation

$$f.t = G_t f \quad ,$$

where G_t is given by Theorem 4.1.

4.3'. Remark

The unicity of the kernel $G_{V_o 1}$ associated to $V_o 1$ shows that

$$f.(V_o 1)=V_o f \quad \text{for each} \quad f\in\mathcal{B}_b(E) \quad .$$

4.4. Lemma

Let U be an open set such that $P^x(X_{T_{CU}} \in E\setminus\overline{U})=0$ for each $x\in U$. Assume that $s,t\in P\cap C_b(E)$ are such that $s=t$ on $\overline{U}$. If $u\in P\cap C_b(E)$ is such that $s-u\in P$ and there exists a compact set, $K\subset U$, such that $P_K u=u$, then $t-u\in P$.

Proof

We are going to apply Theorem 3.2 for the function t-u. If $x\in U$ we put $\mathcal{U}(x)=\{W \text{ open} \mid \overline{W}\subset U,\ x\in W\}$. The condition from the statement implies $P^x(X_{T_{CW}}\in E\setminus\overline{U})=0$, and hence $P_{CW}t(x)=P_{CW}s(x)$ for each $W\in\mathcal{U}(x)$. Then $(t-u)(x)-P_{CW}(t-u)(x)=(s-u)(x)-P_{CW}(s-u)(x)\geq 0$.

If $x\in E\setminus U$ we put $\mathcal{U}(x)=\{W \text{ open} \mid W\cap K=\emptyset,\ x\in W\}$. Since $P_K u(x)=P_{CW}u(x)$, we have

$$(t-u)(x)-P_{CW}(t-u)(x)=t(x)-P_{CW}t(x)\geq 0.$$

Then Theorem 3.2 implies that t-u is nonnegative and excessive. Since $t-u\leq t\in P$ we have $t-u\in P$.

4.5. Proposition

If U, s, t satisfy the requirements of the preceding Lemma, then $f.s=f.t$ for each $f\in\mathcal{B}_b(E)$ which satisfies $f=0$ on $E\setminus U$.

Proof

From the construction of the kernel $\mathcal{G}_s$ (see [30] p.239) it follows that for each open set D,

$\chi_D \cdot s = \sup\{u \in S \cap C(E) \mid s-u \in S$ and $R(u\chi_k)=u$ for some compact set $K \subset D\}$.

A similar relation holds for $\chi_D \cdot t$, and the equality $\chi_D \cdot s = \chi_D \cdot t$ follows from the preceding lemma for $D \subset U$. Further the monotone class theorem implies the desired conclusion.

4.6. Lemma

Let U be an open and $x_o \in U$. Then there exist two functions $p,q \in P \cap C_b(E)$ such that $p \geq q$, $p-q \in C_c(U)$ and $p(x_o) > q(x_o)$.

Proof

Put $p=V_o 1$ and choose a function $g \in C(E)$ such that $g=0$ on an open neighbourhood D of x_o, $0 \leq g \leq 1$ and $g=1$ on $E \setminus U$. Then put $q=R(gp)$. From Theorem 2.2, 3° we get $q \in C(E)$. On the other hand we have

$$P_{CD}p(x_o) = E^{x_o}[\zeta - T_{CD}] < E^{x_o}[\zeta] = p(x_o)$$

From Hunt's theorem (see [6] p.141) it follows $P_{CD}p(x_o) = R(q\chi_{CD})(x_o)$. Since $R(q\chi_{CD}) = q$ we get $q(x_o) < p(x_o)$.

4.7. Lemma

Let u be a continuous excessive function and K a compact set such that $P_K u = u$. If $\{u_n\}$ is an increasing sequence of continuous excessive functions which converges to u, then the convergence is uniform.

Proof

By Dini's theorem we deduce that for each $\varepsilon > 0$ there exists $n \in N$ such that $u \leq u_n + \varepsilon$ on K. Then

$$u = P_K u \leq P_K(u_n + \varepsilon) \leq u_n + \varepsilon \quad \text{on} \quad E.$$

4.8. Proposition

Let u be a continuous excessive function and K a compact set such that $P_K u = u$. Assume that $t \in P \cap C(E)$ and $\{f_n\}$ is a sequence of continuous functions such that the sequence $\{f_n.t\}$ is increasing and $\lim_{n\to\infty} f_n.t=u$. If $g\in C_c(E)$ is such that $0\leq g\leq 1$ and $g=1$ on an open set D with $K\subset D$, then $\lim_{n\to\infty} (gf_n).t=u$ uniform.

Proof

Using Lemma 4.6 we first choose two continuous bounded excessive functions $p,q\in P$ such that $p=q$ on $E\setminus D$ and $p-q\geq 1$ on K. Then we can apply Theorem 3.1' of Chapter II with A=K and $B=E\setminus D$. Thus for each $x\in E$ we have a positive measure μ_x such that

$$s(x)=\mu_x(s-P_K s) \text{ for each } s\in P \text{ which fulfils } P_{E\setminus D}s=s \ .$$

Furthermore $\mu_x(1)<||q||$, for each $x\in E$. Hence

$$||s||\leq c||s-P_K s|| \quad \text{for each } s\in P \text{ which fulfils } P_{E\setminus D}s=s \ .$$

From Lemma 4.7 we know that $f_n.t \to u$ uniform. Then $P_K(f_n.t) \to P_K(u)$ uniform. Since $P_K u=u$ we deduce $f_n.t-P_K(f_n.t) \to 0$. Further from the inequality

$$f_n.t-P_K(f_n.t)=(gf_n).t-P_K((gf_n).t)+((1-g)f_n).t-P_K(((1-g)f_n).t)$$

$$\geq((1-g)f_n).t-P_K(((1-g)f_n).t)\geq(1/c)((1-g)f_n).t$$

we get $((1-g)f_n).t \to 0$, which implies

$$(gf_n).t=f_n.t-((1-g)f_n).t \to u.$$

4.9. Remark. If in the preceding proposition we assume K is closed and CK is relatively compact instead of assuming K is compact, then the conclusion is still valid with uniform convergence on each compact subset of E instead of uniform convergence on the whole space E.

4.10. Theorem

Let U be a relatively compact open set such that for each $x\in CU$, $P^x(X_{T_{\bar U}}\in U)=0$. Then for each open set, A, such that $\bar U\subset A$, the following inclusion holds:

$$C_c(U) \subset \overline{V_o(C_c(A))}$$

Proof

Let us define

$$T(U)=\{f \in C_c(U) \mid \text{there exist } s,t \in P \cap C_b(E) \text{ such that } f=s-t\}$$

From Lemma 4.6 and the Stone-Weierstrass theorem it follows

$$\overline{T(U)}=C_o(U) \ .$$

Therefore for each $f \in C_c(U)$ and each $\varepsilon>0$ there exist $s,t \in P \cap C_b(E)$ such that $s-t \in C_c(U)$ and $|s-t-f|<\varepsilon$. Let now $g \in C_c(A)$ be such that $0 \leq g \leq 1$ and $g=1$ on U. From Lemma 4.5 we deduce $(1-g).s=(1-g).t$ and hence $g.s-g.t=s-t$.

Since g.s is excessive we have $\lim_{\lambda\to\infty} \lambda V_\lambda(g.s)=g.s$. If we put $p=V_o 1$ and $f_n=n(g.s-nV_n(g.s))$ we have $nV_n(g.s)=f_n.p$. Now we apply Proposition 4.8 for K=suppg and u=g.s. Let $g' \in C_c(A)$ be such that $0 \leq g' \leq 1$ and $g'=1$ on a neighbourhood of K. Then there exists $n \in N$ such that $|(g'f_n).p-g.s|<\varepsilon$. Putting $h=g'f_n$ we can writte $(g'f_n).p=h.p=V_o h$ and $|V_o h-g.s|<\varepsilon$.

Similarly we can find $h' \in C_c(A)$ such that $|V_o h'-g.t|<\varepsilon$. Therefore

$$|V_o(h-h')-f|<3\varepsilon \quad \text{and } h-h' \in C_c(A) \ .$$

4.11. Lemma

Let A,U be two open sets such that $U \subset A$. If

$$C_c(U) \subset \overline{V_o(C_c(A))} \quad ,$$

then for each $x \in E \setminus A$ we have $P^x(X_{T_A} \in U)=0$.

Proof

If $f \in C_c(A)$ and $x \in E \setminus A$, then we have $E^x[\int_0^{T_A} f(X_t)dt]=0$, and hence

$$P_A V_o f(x)=E^x[\int_{T_A}^{\infty} f(X_t)dt]=V_o f(x)=0 .$$

The density condition leads to $P_A g(x)=0$ for each $g \in C_c(U)$, which implies $P^x(X_{T_A} \in U)=0$.

4.12. Corollary

The process $(\Omega, M, M_t, X_t, \theta_t, P^x)$ is continuous if and only if for each open set, U, the following inclusion holds:

$$(*) \qquad C_c(U) \subset \overline{V_o(C_c(U))}$$

Proof

If the process is continuous one uses Theorem 4.10 and get relation (*) for each open set.

Now let us suppose that relation (*) is valid for each open set. Let W be an open set and put $U=E\setminus\overline{W}$. From Lemma 4.11 we get

$$P^x(X_{T_{E\setminus W}} \in E\setminus\overline{W})=0 \quad \text{for each} \quad x \in W .$$

Then from Lemma 2.4 of Chapter I we deduce that the process is continuous.

Note on the Product of Semigroups in Hilbert Spaces

Let H be a Hilbert space. If K is a compact set in H, then for each $\varepsilon>0$ there exists a finite dimensional subspace H_o such that the projection operator P_{H_o} satisfies $||x-P_{H_o}x||<\varepsilon$ for each $x\in K$. Therefore if A is a compact operator on H, then A can be approximated uniform with operators of the form $P_{H_o}A$, with H_o a finite dimensional subspace.

We recall that the uniform limit of a sequence of compact operators is also compact. We also recall that a C_o-class semigroup of contractions on H is a family $\{P_t/t>0\}$ of linear operators on H such that $||P_t||\leq 1$, $P_tP_s=P_{t+s}$ and $\lim_{t\to 0} P_tx=x$ for each $x\in H$.

1. Theorem. Let H_1, H_2 be two Hilbert spaces. For i=1,2 let $\{P_t^i|t>0\}$ be a C_o-class semigroup of contractions on H_i. On the tensor product $H_1\otimes H_2$ there is a C_o-class semigroup of contractions $\{P_t,t>0\}$ defined by $P_t=P_t^1\otimes P_t^2$. For $\lambda>0$ let

$$R_\lambda^i=\int_0^\infty \exp(-\lambda t)P_t^i dt\ ,\qquad i=1,2$$

$$R_\lambda=\int_0^\infty \exp(-\lambda t)P_t dt\ .$$

If the operators P_t^1, $t>0$ and R_λ^2, $\lambda>0$ are compact, then the operators R_λ, $\lambda>0$ are also compact.

Proof. In order to prove the theorem it suffices to show that the operators

$$R_{\lambda,a}=\int_{2a}^\infty \exp(-\lambda t)P_t dt\ ,\qquad a>0\ ,\ \lambda>0$$

are compact. Since $R_{\lambda,a}=\exp(-2\lambda a)(I\otimes P_{2a}^2)\circ(P_a^1\otimes I)\circ R_\lambda\circ(P_a^1\otimes I)$ the theorem will be proved if we show that $(P_a^1\otimes I)\circ R_\lambda\circ(P_a^1\otimes I)$ is compact. Further since the operator P_a^1 can be uniform aproximated with operators of the form $P_{H_o} P_a^1$, with H_o a finite dimensional subspace of H_1 , it suffices to show that each operator of the form

$$A=(P_{H_o} P_a^1 \otimes I) o R_\lambda o (P_{H_o} P_a^1 \otimes I)$$

is a compact operator. Let $\{e_1,\ldots,e_n\}$ be an orthonormal base in Ho. Then we have

$$P_{H_o}(x)=\sum_{i=1}^{n} \langle x, e_i\rangle e_i \quad \text{for each} \quad x\in H_1 .$$

The operator A can be written as

$$A=\int_0^\infty \exp(-\lambda t)(P_{H_o} P_a^1 P_t^1 P_{H_o} P_a^1) \otimes P_t^2 dt=$$

$$=\int_0^\infty \exp(-\lambda t)\Big(\sum_{i=1}^{n} \sum_{j=1}^{n} \langle \cdot\, , P_a^{1*} e_j\rangle\langle e_j , P_{t+a}^{1*} e_i\rangle e_i\Big) \otimes P_t^2 .$$

Therefore in order to show that A is compact it suffices to prove that

$$B=\int_0^\infty \exp(-\lambda t)\langle e_j , P_{t+a}^{1*} e_i\rangle P_t^2 dt$$

is a compact operator on H_2.

The function $f(t)=\langle e_j , P_{t+a}^{1*} e_i\rangle$ is bounded on $[0,\infty)$ and the measure $d\mu(t)=\exp(-\lambda t)dt$ is finite. From the Stone-Weierstrass theorem it follows that f can be approximated in $L^1(\mu)$ with function of the form

$$\sum_{i=1}^{p} c_i \exp(-\alpha_i t), \text{ with } \alpha_i>0,\ c_i\in R,\ i=1,\ldots,p.$$

We deduce that B can be approximated with operators of the form

$$\sum_{i=1}^{p} c_i \int_0^\infty \exp(-(\lambda+\alpha_i)t) P_t^2 dt = \sum_{i=1}^{p} c_i R^2_{\lambda+\alpha_i} ,$$

which are compact. This finishes the proof.

Now we are going to give a converse to the above theorem. Let $H_o=L^2(0,1)$ and denote by $\{P_t^o,\ t>0\}$ the semigroup defined by

$$P_t^o f(x)=\bar{f}(x+t), \quad x\in(0,1)$$

where for $f\in L^2(0,1)$ the function $\bar{f}$ will be defined by

$$\bar{f}(x)=\begin{cases} f(x) & \text{if } x\in(0,1) \\ 0 & \text{if } x\notin(0,1) \end{cases}.$$

Obviously $\{P_t^o, t>0\}$ is a C_o-class semigroup of contractions on H_o. Let $R_\lambda^o=\int_0^\infty \exp(-\lambda t)P_t dt$, $\lambda>0$. The potential operator R_o^o is a compact operator. Endeed for $f\in L^2(0,1)$ and $x<y$ we have

$$R_o^o f(x)-R_o^o f(y)=\int_x^y f(t)dt\leq (y-x)^{1/2}||f||_2 \quad .$$

Therefore the operators $\{R_\lambda^o|\lambda>0\}$ are all compact. As a converse to the above theorem we have the following result:

2. Theorem

Let H be a Hilbert space and $\{P_t, t>0\}$ a C_o-class semigroup of operators on H. If there exists $\lambda>0$ such that the operator

$$\int_0^\infty \exp(-\lambda t)P_t^o \otimes P_t dt \quad ,$$

is compact, then the operators P_t, $t>0$ are also compact.

Proof. We shall use the identification $H_o \otimes H=L^2((0,1),H)$. For $x\in H$ we put $f_x(t)=\exp(\lambda t)P_{1-t}x$. Obviously $f_x\in L^2((0,1),H)$. A stightforward computation gives us

$$(\int_0^\infty \exp(-\lambda t)P_t^o \otimes P_t dt f_x)(s)=$$

$$\int_0^{1-s} \exp(-\lambda t)\exp(\lambda s)\exp(\lambda t)P_t P_{1-s-t}x=$$

$$=(1-s)\exp(\lambda s)P_{1-s}x \quad .$$

If we put $g_x(t)=(1-t)\exp(\lambda t)P_{1-t}x$, $0<t<1$, then the assumption of the theorem implies that the set $\{g_x|\,||x||\leq 1\}\subset L^2((0,1),H)$

is compact.

Now we assert that for each $0<s<1$ the set $\{P_{1-s}x \mid ||x||<1\}$ is compact. Endeed let $\{x_n\}$ be a sequence such that $g_{x_n} \to f \in L^q((0,1),H)$ a.s. Then there exists $t \in (s,1)$ such that $g_{x_n}(t) \to f(t)$ in H, and hence $P_{1-t}x_n \to y = \frac{1}{1-t}\exp(-\lambda t)f(t)$. We deduce $P_{1-s}x_n = P_{t-s}P_{1-t}x_n \to P_{t-s}y$. This implies the assertion of the theorem.

Remark. The results of this section are analogous to those proved in Section 5 of Chapter IV, for positive semigroups on function spaces. However it is interesting to note that the kind of tensor product considered there is fundamentally different from the functional point of view.

REFERENCES

[1.] Bauer, H., Harmonische Räume und ihre Potential-theorie, Lecture Notes in Math. 22, 1966.

[2.] Bauer, H., Harmonic spaces and associated Markov processes, in vol.Potential Theory, Roma, Edizioni Cremonese, 1970.

[3.] Berg, C., Potential Theory on the Infinite Dimentional Torus, Inventiones Math., 32, 49-100, 1976.

[4.] Berg, C., Forst, G., Potential Theory on Locally Compact Abelian Groups, Berlin-Heidelberg-New York, Springer, 1975.

[5.] Bliedtner, J., Hansen, W., Markov Processes and Harmonic Spaces, Z. Wahrscheinlichkeistheorie, 42, 309-325, 1978.

[6.] Blumenthal, R.M. Getoor, R.K. Markov Processes and Potential Theory, New York-London, Academic Press, 1968.

[7.] Boboc, N., Bucur, Gh., Convex cones of continuous functions on compact spaces, Editura Academiei, Bucharest, 1976 (Romanian).

[8.] Boboc, N., Constantinescu, C., Cornea, A., Semigroups of transitions on harmonic spaces, Rev.Roum.Math. 12, 763-805, 1967.

[9.] Bony, J.M., Opérateurs elliptiques dégénérés associés aux axiomatiques de la théorie du potentiel, in vol. Potential Theory, Edizioni Cremonese, Roma, 1970.

[10.] Brelot, Lectures on Potential Theory, Tata Inst. of F.R., Bombay 1960 (reissued 1967).

[11.] Cairoli, R., Produits de semi-groupes de transition et produits de processus, Publ.Inst.Stat.Univ. Paris, 4, 1966.

[12.] Cairoli, R., Une théorème sur les fonctions séparement excessives, C.R.A.S., Paris, 263, 161-163, 1967.

[13.] Constantinescu, C., Cornea, A., Potential Theory on Harmonic Spaces, Springer, Berlin-Heidelberg-New York, 1972.

[14.] Courrège, Ph., Priouret, P., Axiomatique du problème de Dirichlet et processus de Markov, Seminaire Brelot-Choquet-Deny, 1963-1964.

[15.] Courrège, Ph., Priouret, P., Recollements de processus de Markov, Publ.Inst.Stat.Univ.Paris, 14, 275-377, 1965.

[16.] Cuculescu, I., Markov Processes and excessive functions, Editura Academiei, Bucharest, 1968 (Romanian).

[17.] Dynkin, E.B., Markov Processes I and II, Springer, Berlin--Götingen-Heidelberg, 1965.

[18.] Gowrisankaran, K., Limites fines et fonctions doublement harmoniques, C.R.A.S., Paris, 262, 388-330, 1966.

[19.] Hansen, W., Konstruction von Halbgruppen und Markoffschen

Prozessen, Inventiones Math., 3, 179-214, 1967.

[20.] Hervé, R-M., Developpements sur une théorie axiomatique des fonctions surharmoniques, C.R.A.S. Paris 248, 179-181, 1959.

[21.] Hervé, R-M., Recherches axiomatiques sur la théorie des fonctions surharmonique et du potentiel, Ann.Inst.Fourier, 12, 415-571, 1962.

[22.] Hirsch, F., Familles résolvantes, générateurs, cogénérateurs, poentiels, Ann.Inst.Fourier, 22, 89-210, 1972.

[23.] Lumer, G., Problème de Cauchy pour opérateurs locauy et "changement du temp", Ann.Inst.Fourier, 25, 409-446, 1975.

[24.] Lumer, G., (I) Problème du Cauchy avec valeur au bord continues, comportement asymptotique et applications, (II) Problème de Cauchy et fonctions surharmoniques, Seminaire de Théorie du Potentiel no.2, Lecture Notes Math. 563, Berlin-Heidelberg--New York, Springer, 1976.

[26.] Meyer, P.A., Brelot's axiomatic theory of the Dirichlet problem and Hunt's theory, Ann.Inst.Fourier, 13, 357-372, 1963.

[27.] Meyer, P.A., Probability and Potentials, Waltham-Massachusets--Toronto-London, Blaisdell Publishing Company, 1966.

[28.] Meyer, P.A., Les résolvantes fortement felleriennes, d'après Mokobodzki, Seminaire de Probabilités II, Lecture Notes in Math. 51, 1968.

[29.] Mokobodzki, G., Densité relative de deux potentiels comparables, sans ultrafiltre rapide, Séminaire de Probabilités IV, Lecture Notes Math. 124, 1970.

[30.] Mokobodzki, G., Cônes de potentiels et noyaux subordonés, in vol. Potential Theory, Edizione Cremonese, Roma, 1970.

[31.] Mokobodzki, G., Sibony, D., Familles additives de cônes convexes et noyaux subordonnés. Ann.Inst.Fourier, 18, 205-220, 1969.

[32.] Oleinik, O.A., Radkevich, E.B., Second order equations with non-negative characteristic form, Itogy Nauky, Moskow, 1971 (Russian).

[33.] Popa, E., Thesis, University of Bucharest, 1978.

[34.] Roth, J-P., Opérateurs dissipatifs et semi-groupes dans les espaces de fonctions continues, Ann.Inst. Fourier, 26, 1-98, 1976.

[35.] Roth, J.P., Les operateurs elliptiques comme generateurs infinitesimaux de semi-groupes de Feller. Séminaire de Théorie du Potentiel nr.3, Lecture Notes in Math. 681, 1978.

[36.] Schirmeier, U., Produkte harmonischer Räume, Bayerische Akademie Wiss., München, 1978.

[37.] Stoica, L., On the construction of a kernel, Rev.Roum.Math. 22

fasc.8, p.1167-1172, 1977.

[38.] Stoica, L., Thesis, University of Bucharest, 1978.

[39.] Stoica, L., On excessive functions, Proc.Japan Acad. 55, A, no.3, 85-87, 1979.

[40.] Taylor, J.C., The Harmonic Space Associated with a "Reasonable" Standard Process,Math.Ann. 233, 89-96, 1978.

[41.] Watanabe, T., On the equivalence of excessive functions and superharmonic functions in the theory of Markov processes, I and II, Proc.Japan Acad.38, 397-401, 402-407, 1962.

[42.] Meyer, P.A. Renaissance, recollements, melanges, ralentissement de processus de Markov, Ann.Inst. Fourier, 25, 465-498, 1975.

[43.] Nagasawa, M., Note on Pasting of Two Markov Processes, Séminaire de Probabilités X, Lecture Notes Math. 511, Springer 1976.

[44.] Bliedtner, J., Harmonische Gruppen und Huntsche Faltungskerne, Seminar über Potentialtheorie, 69-102, Lecture Notes in Mathematics 69, Springer 1968.

[45.] Taylor, J.C.,Ray Processes on Locally Compact Spaces, Math.Ann.208, 233-248, 1974.

[46.] Schirmeier, U., Konvergenzeigenschaften in harmonischen Räumen, Inventiones Mathematicae 55, 71-95, 1979.

Subject Index

List of Symbols

Vol. 640: J. L. Dupont, Curvature and Characteristic Classes. X, 175 pages. 1978.

Vol. 641: Séminaire d'Algèbre Paul Dubreil, Proceedings Paris 1976–1977. Edité par M. P. Malliavin. IV, 367 pages. 1978.

Vol. 642: Theory and Applications of Graphs, Proceedings, Michigan 1976. Edited by Y. Alavi and D. R. Lick. XIV, 635 pages. 1978.

Vol. 643: M. Davis, Multiaxial Actions on Manifolds. VI, 141 pages. 1978.

Vol. 644: Vector Space Measures and Applications I, Proceedings 1977. Edited by R. M. Aron and S. Dineen. VIII, 451 pages. 1978.

Vol. 645: Vector Space Measures and Applications II, Proceedings 1977. Edited by R. M. Aron and S. Dineen. VIII, 218 pages. 1978.

Vol. 646: O. Tammi, Extremum Problems for Bounded Univalent Functions. VIII, 313 pages. 1978.

Vol. 647: L. J. Ratliff, Jr., Chain Conjectures in Ring Theory. VIII, 133 pages. 1978.

Vol. 648: Nonlinear Partial Differential Equations and Applications, Proceedings, Indiana 1976–1977. Edited by J. M. Chadam. VI, 206 pages. 1978.

Vol. 649: Séminaire de Probabilités XII, Proceedings, Strasbourg, 1976–1977. Edité par C. Dellacherie, P. A. Meyer et M. Weil. VIII, 805 pages. 1978.

Vol. 650: C*-Algebras and Applications to Physics. Proceedings 1977. Edited by H. Araki and R. V. Kadison. V, 192 pages. 1978.

Vol. 651: P. W. Michor, Functors and Categories of Banach Spaces. VI, 99 pages. 1978.

Vol. 652: Differential Topology, Foliations and Gelfand-Fuks-Cohomology, Proceedings 1976. Edited by P. A. Schweitzer. XIV, 252 pages. 1978.

Vol. 653: Locally Interacting Systems and Their Application in Biology. Proceedings, 1976. Edited by R. L. Dobrushin, V. I. Kryukov and A. L. Toom. XI, 202 pages. 1978.

Vol. 654: J. P. Buhler, Icosahedral Golois Representations. III, 143 pages. 1978.

Vol. 655: R. Baeza, Quadratic Forms Over Semilocal Rings. VI, 199 pages. 1978.

Vol. 656: Probability Theory on Vector Spaces. Proceedings, 1977. Edited by A. Weron. VIII, 274 pages. 1978.

Vol. 657: Geometric Applications of Homotopy Theory I, Proceedings 1977. Edited by M. G. Barratt and M. E. Mahowald. VIII, 459 pages. 1978.

Vol. 658: Geometric Applications of Homotopy Theory II, Proceedings 1977. Edited by M. G. Barratt and M. E. Mahowald. VIII, 487 pages. 1978.

Vol. 659: Bruckner, Differentiation of Real Functions. X, 247 pages. 1978.

Vol. 660: Equations aux Dérivée Partielles. Proceedings, 1977. Edité par Pham The Lai. VI, 216 pages. 1978.

Vol. 661: P. T. Johnstone, R. Paré, R. D. Rosebrugh, D. Schumacher, R. J. Wood, and G. C. Wraith, Indexed Categories and Their Applications. VII, 260 pages. 1978.

Vol. 662: Akin, The Metric Theory of Banach Manifolds. XIX, 306 pages. 1978.

Vol. 663: J. F. Berglund, H. D. Junghenn, P. Milnes, Compact Right Topological Semigroups and Generalizations of Almost Periodicity. X, 243 pages. 1978.

Vol. 664: Algebraic and Geometric Topology, Proceedings, 1977. Edited by K. C. Millett. XI, 240 pages. 1978.

Vol. 665: Journées d'Analyse Non Linéaire. Proceedings, 1977. Edité par P. Bénilan et J. Robert. VIII, 256 pages. 1978.

Vol. 666: B. Beauzamy, Espaces d'Interpolation Réels: Topologie et Géometrie. X, 104 pages. 1978.

Vol. 667: J. Gilewicz, Approximants de Padé. XIV, 511 pages. 1978.

Vol. 668: The Structure of Attractors in Dynamical Systems. Proceedings, 1977. Edited by J. C. Martin, N. G. Markley and W. Perrizo. VI, 264 pages. 1978.

Vol. 669: Higher Set Theory. Proceedings, 1977. Edited by G. H. Müller and D. S. Scott. XII, 476 pages. 1978.

Vol. 670: Fonctions de Plusieurs Variables Complexes III, Proceedings, 1977. Edité par F. Norguet. XII, 394 pages. 1978.

Vol. 671: R. T. Smythe and J. C. Wierman, First-Passage Perculation on the Square Lattice. VIII, 196 pages. 1978.

Vol. 672: R. L. Taylor, Stochastic Convergence of Weighted Sums of Random Elements in Linear Spaces. VII, 216 pages. 1978.

Vol. 673: Algebraic Topology, Proceedings 1977. Edited by P. Hoffman, R. Piccinini and D. Sjerve. VI, 278 pages. 1978.

Vol. 674: Z. Fiedorowicz and S. Priddy, Homology of Classical Groups Over Finite Fields and Their Associated Infinite Loop Spaces. VI, 434 pages. 1978.

Vol. 675: J. Galambos and S. Kotz, Characterizations of Probability Distributions. VIII, 169 pages. 1978.

Vol. 676: Differential Geometrical Methods in Mathematical Physics II, Proceedings, 1977. Edited by K. Bleuler, H. R. Petry and A. Reetz. VI, 626 pages. 1978.

Vol. 677: Séminaire Bourbaki, vol. 1976/77, Exposés 489–506. IV, 264 pages. 1978.

Vol. 678: D. Dacunha-Castelle, H. Heyer et B. Roynette. Ecole d'Eté de Probabilités de Saint-Flour. VII-1977. Edité par P. L. Hennequin. IX, 379 pages. 1978.

Vol. 679: Numerical Treatment of Differential Equations in Applications, Proceedings, 1977. Edited by R. Ansorge and W. Törnig. IX, 163 pages. 1978.

Vol. 680: Mathematical Control Theory, Proceedings, 1977. Edited by W. A. Coppel. IX, 257 pages. 1978.

Vol. 681: Séminaire de Théorie du Potentiel Paris, No. 3, Directeurs: M. Brelot, G. Choquet et J. Deny. Rédacteurs: F. Hirsch et G. Mokobodzki. VII, 294 pages. 1978.

Vol. 682: G. D. James, The Representation Theory of the Symmetric Groups. V, 156 pages. 1978.

Vol. 683: Variétés Analytiques Compactes, Proceedings, 1977. Edité par Y. Hervier et A. Hirschowitz. V, 248 pages. 1978.

Vol. 684: E. E. Rosinger, Distributions and Nonlinear Partial Differential Equations. XI, 146 pages. 1978.

Vol. 685: Knot Theory, Proceedings, 1977. Edited by J. C. Hausmann. VII, 311 pages. 1978.

Vol. 686: Combinatorial Mathematics, Proceedings, 1977. Edited by D. A. Holton and J. Seberry. IX, 353 pages. 1978.

Vol. 687: Algebraic Geometry, Proceedings, 1977. Edited by L. D. Olson. V, 244 pages. 1978.

Vol. 688: J. Dydak and J. Segal, Shape Theory. VI, 150 pages. 1978.

Vol. 689: Cabal Seminar 76–77, Proceedings, 1976–77. Edited by A.S. Kechris and Y. N. Moschovakis. V, 282 pages. 1978.

Vol. 690: W. J. J. Rey, Robust Statistical Methods. VI, 128 pages. 1978.

Vol. 691: G. Viennot, Algèbres de Lie Libres et Monoïdes Libres. III, 124 pages. 1978.

Vol. 692: T. Husain and S. M. Khaleelulla, Barrelledness in Topological and Ordered Vector Spaces. IX, 258 pages. 1978.

Vol. 693: Hilbert Space Operators, Proceedings, 1977. Edited by J. M. Bachar Jr. and D. W. Hadwin. VIII, 184 pages. 1978.

Vol. 694: Séminaire Pierre Lelong – Henri Skoda (Analyse) Année 1976/77. VII, 334 pages. 1978.

Vol. 695: Measure Theory Applications to Stochastic Analysis, Proceedings, 1977. Edited by G. Kallianpur and D. Kölzow. XII, 261 pages. 1978.

Vol. 696: P. J. Feinsilver, Special Functions, Probability Semigroups, and Hamiltonian Flows. VI, 112 pages. 1978.

Vol. 697: Topics in Algebra, Proceedings, 1978. Edited by M. F. Newman. XI, 229 pages. 1978.

Vol. 698: E. Grosswald, Bessel Polynomials. XIV, 182 pages. 1978.

Vol. 699: R. E. Greene and H.-H. Wu, Function Theory on Manifolds Which Possess a Pole. III, 215 pages. 1979.

Vol. 700: Module Theory, Proceedings, 1977. Edited by C. Faith and S. Wiegand. X, 239 pages. 1979.

Vol. 701: Functional Analysis Methods in Numerical Analysis, Proceedings, 1977. Edited by M. Zuhair Nashed. VII, 333 pages. 1979.

Vol. 702: Yuri N. Bibikov, Local Theory of Nonlinear Analytic Ordinary Differential Equations. IX, 147 pages. 1979.

Vol. 703: Equadiff IV, Proceedings, 1977. Edited by J. Fábera. XIX, 441 pages. 1979.

Vol. 704: Computing Methods in Applied Sciences and Engineering, 1977, I. Proceedings, 1977. Edited by R. Glowinski and J. L. Lions. VI, 391 pages. 1979.

Vol. 705: O. Forster und K. Knorr, Konstruktion verseller Familien kompakter komplexer Räume. VII, 141 Seiten. 1979.

Vol. 706: Probability Measures on Groups, Proceedings, 1978. Edited by H. Heyer. XIII, 348 pages. 1979.

Vol. 707: R. Zielke, Discontinuous Čebyšev Systems. VI, 111 pages. 1979.

Vol. 708: J. P. Jouanolou, Equations de Pfaff algébriques. V, 255 pages. 1979.

Vol. 709: Probability in Banach Spaces II. Proceedings, 1978. Edited by A. Beck. V, 205 pages. 1979.

Vol. 710: Séminaire Bourbaki vol. 1977/78, Exposés 507–524. IV, 328 pages. 1979.

Vol. 711: Asymptotic Analysis. Edited by F. Verhulst. V, 240 pages. 1979.

Vol. 712: Equations Différentielles et Systèmes de Pfaff dans le Champ Complexe. Edité par R. Gérard et J.-P. Ramis. V, 364 pages. 1979.

Vol. 713: Séminaire de Théorie du Potentiel, Paris No. 4. Edité par F. Hirsch et G. Mokobodzki. VII, 281 pages. 1979.

Vol. 714: J. Jacod, Calcul Stochastique et Problèmes de Martingales. X, 539 pages. 1979.

Vol. 715: Inder Bir S. Passi, Group Rings and Their Augmentation Ideals. VI, 137 pages. 1979.

Vol. 716: M. A. Scheunert, The Theory of Lie Superalgebras. X, 271 pages. 1979.

Vol. 717: Grosser, Bidualräume und Vervollständigungen von Banachmoduln. III, 209 pages. 1979.

Vol. 718: J. Ferrante and C. W. Rackoff, The Computational Complexity of Logical Theories. X, 243 pages. 1979.

Vol. 719: Categorial Topology, Proceedings, 1978. Edited by H. Herrlich and G. Preuß. XII, 420 pages. 1979.

Vol. 720: E. Dubinsky, The Structure of Nuclear Fréchet Spaces. V, 187 pages. 1979.

Vol. 721: Séminaire de Probabilités XIII. Proceedings, Strasbourg, 1977/78. Edité par C. Dellacherie, P. A. Meyer et M. Weil. VII, 647 pages. 1979.

Vol. 722: Topology of Low-Dimensional Manifolds. Proceedings, 1977. Edited by R. Fenn. VI, 154 pages. 1979.

Vol. 723: W. Brandal, Commutative Rings whose Finitely Generated Modules Decompose. II, 116 pages. 1979.

Vol. 724: D. Griffeath, Additive and Cancellative Interacting Particle Systems. V, 108 pages. 1979.

Vol. 725: Algèbres d'Opérateurs. Proceedings, 1978. Edité par P. de la Harpe. VII, 309 pages. 1979.

Vol. 726: Y.-C. Wong, Schwartz Spaces, Nuclear Spaces and Tensor Products. VI, 418 pages. 1979.

Vol. 727: Y. Saito, Spectral Representations for Schrödinger Operators With Long-Range Potentials. V, 149 pages. 1979.

Vol. 728: Non-Commutative Harmonic Analysis. Proceedings, 1978. Edited by J. Carmona and M. Vergne. V, 244 pages. 1979.

Vol. 729: Ergodic Theory. Proceedings, 1978. Edited by M. Denker and K. Jacobs. XII, 209 pages. 1979.

Vol. 730: Functional Differential Equations and Approximation of Fixed Points. Proceedings, 1978. Edited by H.-O. Peitgen and H.-O. Walther. XV, 503 pages. 1979.

Vol. 731: Y. Nakagami and M. Takesaki, Duality for Crossed Products of von Neumann Algebras. IX, 139 pages. 1979.

Vol. 732: Algebraic Geometry. Proceedings, 1978. Edited by K. Lønsted. IV, 658 pages. 1979.

Vol. 733: F. Bloom, Modern Differential Geometric Techniques in the Theory of Continuous Distributions of Dislocations. XII, 206 pages. 1979.

Vol. 734: Ring Theory, Waterloo, 1978. Proceedings, 1978. Edited by D. Handelman and J. Lawrence. XI, 352 pages. 1979.

Vol. 735: B. Aupetit, Propriétés Spectrales des Algèbres de Banach. XII, 192 pages. 1979.

Vol. 736: E. Behrends, M-Structure and the Banach-Stone Theorem. X, 217 pages. 1979.

Vol. 737: Volterra Equations. Proceedings 1978. Edited by S.-O. Londen and O. J. Staffans. VIII, 314 pages. 1979.

Vol. 738: P. E. Conner, Differentiable Periodic Maps. 2nd edition, IV, 181 pages. 1979.

Vol. 739: Analyse Harmonique sur les Groupes de Lie II. Proceedings, 1976–78. Edited by P. Eymard et al. VI, 646 pages. 1979.

Vol. 740: Séminaire d'Algèbre Paul Dubreil. Proceedings, 1977–78. Edited by M.-P. Malliavin. V, 456 pages. 1979.

Vol. 741: Algebraic Topology, Waterloo 1978. Proceedings. Edited by P. Hoffman and V. Snaith. XI, 655 pages. 1979.

Vol. 742: K. Clancey, Seminormal Operators. VII, 125 pages. 1979.

Vol. 743: Romanian-Finnish Seminar on Complex Analysis. Proceedings, 1976. Edited by C. Andreian Cazacu et al. XVI, 713 pages. 1979.

Vol. 744: I. Reiner and K. W. Roggenkamp, Integral Representations. VIII, 275 pages. 1979.

Vol. 745: D. K. Haley, Equational Compactness in Rings. III, 167 pages. 1979.

Vol. 746: P. Hoffman, τ-Rings and Wreath Product Representations. V, 148 pages. 1979.

Vol. 747: Complex Analysis, Joensuu 1978. Proceedings, 1978. Edited by I. Laine, O. Lehto and T. Sorvali. XV, 450 pages. 1979.

Vol. 748: Combinatorial Mathematics VI. Proceedings, 1978. Edited by A. F. Horadam and W. D. Wallis. IX, 206 pages. 1979.

Vol. 749: V. Girault and P.-A. Raviart, Finite Element Approximation of the Navier-Stokes Equations. VII, 200 pages. 1979.

Vol. 750: J. C. Jantzen, Moduln mit einem höchsten Gewicht. III, 195 Seiten. 1979.

Vol. 751: Number Theory, Carbondale 1979. Proceedings. Edited by M. B. Nathanson. V, 342 pages. 1979.

Vol. 752: M. Barr, *-Autonomous Categories. VI, 140 pages. 1979.

Vol. 753: Applications of Sheaves. Proceedings, 1977. Edited by M. Fourman, C. Mulvey and D. Scott. XIV, 779 pages. 1979.

Vol. 754: O. A. Laudal, Formal Moduli of Algebraic Structures. III, 161 pages. 1979.

Vol. 755: Global Analysis. Proceedings, 1978. Edited by M. Grmela and J. E. Marsden. VII, 377 pages. 1979.

Vol. 756: H. O. Cordes, Elliptic Pseudo-Differential Operators – An Abstract Theory. IX, 331 pages. 1979.

Vol. 757: Smoothing Techniques for Curve Estimation. Proceedings, 1979. Edited by Th. Gasser and M. Rosenblatt. V, 245 pages. 1979.

Vol. 758: C. Năstăsescu and F. Van Oystaeyen; Graded and Filtered Rings and Modules. X, 148 pages. 1979.